# Applied ethology 2011:

## Scientific evaluation of behavior, welfare and enrichment

# Proceedings of the
# 45th Congress of the International Society for Applied Ethology (ISAE)

## Scientific evaluation of behavior, welfare and enrichment

*Hyatt Regency Hotel, Indianapolis, U.S.A.*
*31 July – 4 August 2011*

edited by:
Edmond A. Pajor
Jeremy N. Marchant-Forde

*Wageningen Academic*
*P u b l i s h e r s*

ISBN: 978-90-8686-179-8
e-ISBN: 978-90-8686-737-0
DOI: 10.3921/978-90-8686-737-0

First published, 2011

© Wageningen Academic Publishers
The Netherlands, 2011

The individual contributions in this publication and any liabilities arising from them remain the responsibility of the authors.

# Welcome to the 45<sup>th</sup> Congress of the ISAE

On behalf of the Organizing and Scientific Committees, we would like to extend a warm welcome to you during your visit to Indianapolis and the U.S.A. We are very happy to be hosting the Congress for its third visit to the U.S.A. and it is an exciting time for the field of applied animal behavior and welfare in this country. We are seeing an upsurge in public interest in animal welfare and a State-by-State introduction of welfare legislation, primarily aimed at farm animal housing systems. Zoo, laboratory and companion animal welfare issues are also receiving increasing attention, and our scientific program also contains a broad representation of studies of animals in these settings.

The study of animal behavior and welfare has, by U.S. standards, a relatively long history at Purdue University where Animal Sciences, Veterinary Medicine and USDA-ARS researchers work side-by-side. There are good links with faculty at other universities based here in the Midwest and the membership of the Organizing and Scientific Committees are testament to the productive relationships that exist between scientists at these institutions. Historically, the primary focus in such an agriculture-rich environment was farm animals. As we build towards a critical mass of researchers, the breadth of our interests has expanded and now encompasses all areas, in research, teaching and outreach.

However, there is still more that can be done. The central theme of our meeting is the Scientific Evaluation of Behavior, Welfare and Enrichment. Sound science has to be a critical part of the work that we do, to garner the trust of our stakeholders and to genuinely impact the welfare of animals within our care. Meetings such as this offer important opportunities to share knowledge and expertise, across species and across disciplines and we hope you will make the most of the chance to interact with others and learn something new to take home and apply to your work.

As we approach the end of a nearly three-year journey, we would like to take this opportunity to thank the members of the Committees, our Sponsors and also the leadership of the USDA-ARS Livestock Behavior Research Unit and our academic Departments. Time is a precious commodity in all aspects of our life and Departmental support has been instrumental in enabling us to host this Congress.

*Jeremy Marchant-Forde* (USDA-ARS, LBRU – Organizing Committee Chair)
*Ed Pajor* (University of Calgary – Scientific Committee Chair)

# Acknowledgements

## Congress Organizing Committee

Jeremy Marchant-Forde (Chair)
Alan Beck
Candace Croney
Jerry Davis
Joe Garner
Brianna Gaskill

Angela Green
Anna Johnson
Don Lay
Janice Siegford
Janice Swanson
Jason Watters

## Scientific Committee

Ed Pajor (Chair)
Heng-wei Cheng
Candace Croney
Jerry Davis
Susan Eicher

Marcia Endres
Jeremy Marchant-Forde
Suzanne Millman
Janice Siegford

## Ethics Committee

Ian Duncan (Chair)
Stine Christiansen
Maria Jose Hotzel
Don Lay

Francois Martin
Anna Olsson
Alexandra Whittaker

## Professional Conference Organizers

Purdue University Conference Division
Geni Greiner
Stephanie Botkin

Sandra Carter
Kaitlin Misenheimer

## Design

**Congress logo:**
**Congress website:**
**Proceedings cover:**
**Cover photo:**

Jeremy Marchant-Forde
Corey Mann
Wageningen Academic Publishers
Brianna Gaskill

# Referees

Leena Anil
Mike Appleby
Greg Archer
Clover Bench
Harry Blokhuis
Mollie Bloomsmith
John Bradshaw
Elisabetta Canali
Sylvie Cloutier
Mike Cockram
Jonathan Cooper
Rachel Dennis
Trevor Devries
Monica Elmore
Brianna Gaskill
Christy Goldhawk
Derek Haley
Cami Heleski
Laura Hänninen
Margit Bak Jensen
Vanessa Kanaan
Larry Katz
Yuzhi Li
Sue McDonnell
Ruth Newberry
Christine Nicol
Lee Niel

Keelin O'Driscoll
Anne Marie de Passillé
Emily Patterson-Kane
Jose Peralta
Carol Petherick
Anthony Podberscek
Fiona Rioja-Lang
Irene Rochlitz
Steve Ross
Kirsti Rouvinen-Watt
Hans Spoolder
Andreas Steiger
Joe Stookey
Ray Stricklin
Carolyn Stull
Collette Thogerson
Stephanie Torrey
Cassandra Tucker
Frank Tuyttens
Elsa Vasseur
Isabelle Veissier
Kristen Walker
Dan Weary
Bruce Webster
Françoise Wemelsfelder
Hanno Würbel

# Congress Assistants

Shelly Pfeffer Deboer
Melissa Elischer
Brianna Gaskill
Kim McMunn

Jean-Loup Rault
Collette Thogerson
Giovana Viera
Stephanie Wisdom

# Sponsors

**Meeting tomorrow's challenges today**

Agricultural Research Service
www.ars.usda.gov

**Location:** West Lafayette, IN; on the Purdue University Campus

**Research Leader:** Dr. Donald C. Lay

**Area Director:** Dr. Larry Chandler

**National Program 101:** Food Animal Production

**Human Capital:**
5 Scientists representing 5 disciplines: ethology, neuroscience, stress physiology, immunology, and bacteriology.
1 Research Associate
3 Permanent Technicians

www.ars.usda.gov/mwa/lafayette/lbru

**Laboratory Space:** Occupies 17,000 ft$^2$ of laboratory space and 1,800 ft$^2$ of office space.

**Research Focus:** Animal welfare for swine, poultry and dairy cattle; and Pre-harvest Food Safety for swine.

**The Scientific Team:**
- Dr. Donald C. Lay Jr., Ph.D.; Stress Physiologist/Ethologist
- Dr. Susan D. Eicher, Ph.D.; Immunologist
- Dr. Heng wei Cheng, M.D., Ph.D.; Neuroscientist
- Dr. Jeremy N. Marchant-Forde, Ph.D.; Ethologist
- Dr. Marcos H. Rostagno, D.V.M., Ph.D.; Bacteriologist

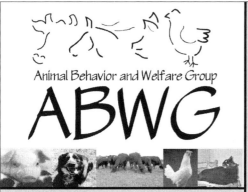

# Venue Maps

## Indianapolis Street Map

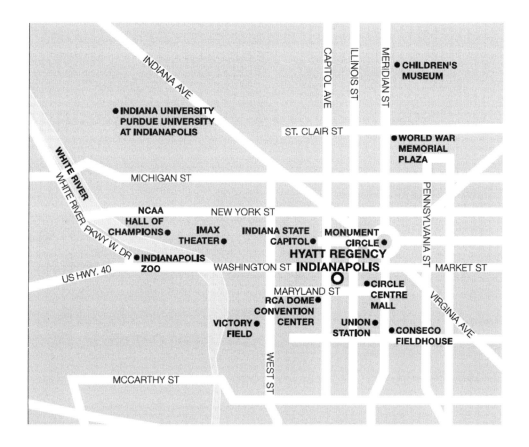

# Hyatt Regency Floor Plans

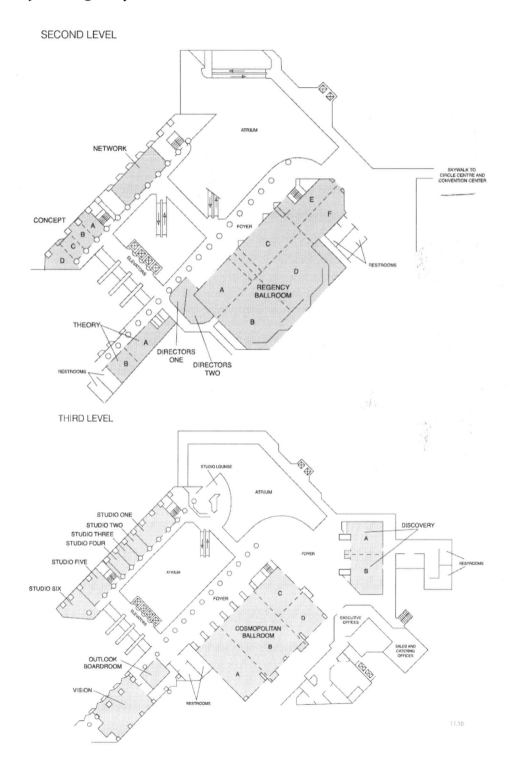

SECOND LEVEL

THIRD LEVEL

# General information

## Conference venue

The Hyatt Regency Indianapolis is located in the Indianapolis downtown area, next to the State Capitol. The address is 1 South Capitol Avenue, Indianapolis, IN 46204.

## Official language

Official language of the meeting is English.

## Registration and information desk

Second Level
**Opening hours:**

| | |
|---|---|
| Sunday July 31: | 12:00 - 9:00 |
| Monday August 1: | 8:00 - 6:00 |
| Tuesday August 2: | 8:00 - 1:30 |
| Wednesday August 3: | 8:00 - 7:00 |
| Thursday August 4: | 8:30 - 6:00 |

E-mail: botkin@purdue.edu

## Name badges

Your name badge is your admission to the venues, scientific sessions, poster sessions and to the lunch and coffee breaks. It should be worn at all times at the conference venue and at social events.

## Poster and exhibition area

The poster and exhibition area will be on the second level of the Hyatt Regency. We encourage you to visit our sponsors.

## Internet access

Wireless Internet access is available will be available in all common meeting space areas, free of charge.

## Receipt of payment and certificate of attendance

If you need a receipt of payment or a certificate of attendance, please ask for it at the registration and information desk.

## Coffee Breaks and Lunches

Coffee and refreshments and Lunch will be served on the second and third levels.

## Welcome reception, July 31st 6:30 - 9:00

The welcome reception will take place at Eiteljorg Museum. Light refreshments and drinks will be served. The venue is within easy walking distance of the hotel, but a minibus shuttle will be also available if weather is bad.

## Excursions, August 5th, 12:30 - 6.00
(Please note that pre-registration is required).

All excursions buses will depart in front of the Hyatt Regency Downtown at 12:00. Lunch is included either en route or at the venue.

## Banquet at Hyatt Regency, August 3rd 7.00-11.00
(Please note that pre-registration is required).

The conference banquet will be held at the Hyatt Regency.

## Farewell party Thursday, August 4th, 5.30 – 10.30

The Farewell Party will be at Victory Field, home of Indianapolis Indians Triple-A baseball team. Gates open at 5.30 pm, game starts at 7.00 pm. Casual dinner and drinks will be served and the party is included in the cost of your registration. An easy walk from the hotel, buses will be available to shuttle you to and from the venue if required.

## Banking service, currency

US dollar ($) is the official currency in the United States. An exchange office is available at Indianapolis Airport by baggage claim (Travelex). There are plenty of cash dispensers (ATMs) in Indianapolis, including inside the Hyatt Regency. Major international credit cards are accepted in most hotels, shops and restaurants.

## Shopping in Indianapolis

The Hyatt Regency is next to Circle Mall in downtown Indianapolis, accessible from the hotel by skywalk, which has more than 100 shopping, dining and entertainment options. The mall is open 10:00 - 9:00 on Monday to Saturday and 10:00 - 6:00 on Sundays. Other stores in Indianapolis have similar opening hours.

## Transport from and to Indianapolis International Airport

*Shuttle*
Carey Transportation (317) 241-7100 or www.careyindiana.com/
From Airport to Downtown Indy (Hyatt): Rate of "Share A Ride" (pickup at anytime) is $16/person plus gratuity. They fill their shuttles up on first come first serve basis. They can have a private 6 passenger vehicle for $62.99 plus gratuity or an 8 passenger vehicle for $68.84 plus gratuity with reservations beforehand.
See www.careyindiana.com/shared_ride.asp for details

*City Bus*
Service to airport
IndyGo's Green Line **Downtown / Airport Express route** provides non-stop service from the airport to convenient locations near major downtown hotels and the Indiana Convention Center. Green Line service runs daily from 5 a.m. to 9 p.m. Cost is $7 per trip.
See www.indygo.net/pages/green-line-downtownairport-express

## Emergency calls

You should call 911 if anything happens which means that a paramedic, the police or the fire department need to be called out.

## Local conference secretariat

Purdue Conference Division
Stewart Center Room 116
128 Memorial Mall
West Lafayette, IN 47907, USA
Tel: +1 800-359-2968

# Program at a glance

| Sun 31st July | | |
|---|---|---|
| 2.00 | Registration Open | |
| 6.30 | Welcome reception – Eiteljorg Museum | |

| Mon 1st Aug | Theater 1 | Theater 2 |
|---|---|---|
| 8.00 | Registration Open | |
| 9.00 | Opening Ceremony | |
| 9.30 | Plenary 1 | |
| 10.00 | Coffee Break & Posters | |
| 10.45 | Pain, Distress & Humane End-Points 1 | Companion Animal Behavior & Welfare |
| 12.00 | Lunch | |
| 1.30 | Plenary 2 | |
| 2.00 | Pain, Distress & Humane End-Points 2 | Maternal Behavior & Effects |
| 3.00 | Coffee Break & Posters | |
| 3.45 | Pain, Distress & Humane End-Points 3 | Communication & Cognition |
| 5.00 | Finish | |

| Tues 2nd Aug | Theater 1 | Theater 2 |
|---|---|---|
| 8.30 | Engineering Environments & Measurement Technologies | Qualitative Behavioral Assessment |
| 9.30 | | Poultry Behavior & Welfare 1 |
| 10.15 | Coffee Break & Posters | |
| 11.00-12.00 | Wood-Gush Memorial Lecture | |
| 12.00- 6.00 | Lunch & Excursions | |

| Wed 3rd Aug | Theater 1 | Theater 2 |
|---|---|---|
| 9.00 | Plenary 3 | |
| 9.30 | Laboratory Animal Behavior, Welfare & Enrichment 1 | Feeding Behavior & Welfare 1 |
| 10.15 | Coffee Break & Posters | |
| 11.00 | Laboratory Animal Behavior, Welfare & Enrichment 2 | Feeding Behavior & Welfare 2 |
| 12.00 | Lunch | |
| 1.30 | Plenary 4 | |
| 2.00 | Laboratory Animal Behavior, Welfare & Enrichment 3 | Sow Housing |
| 3.00 | Coffee Break & Posters | |
| 3.30 | ISAE Annual General Meeting | |
| 5.30 | Finish | |
| 7.00 | Congress Banquet at Hyatt Regency Hotel | |

| Thurs 4th Aug | Theater 1 | Theater 2 |
|---|---|---|
| 9.00 | Plenary 5 | |
| 9.30 | Zoo Animal Behavior, Welfare & Enrichment 1 | Dairy Behavior & Welfare 1 |
| 10.15 | Coffee Break & Posters | |
| 11.00 | Zoo Animal Behavior, Welfare & Enrichment 2 | Human-Animal Interactions |
| 12.00 | Lunch | |
| 1.30 | Plenary 6 | |
| 2.00 | Zoo Animal Behavior, Welfare & Enrichment 3 | Free Papers |
| 3.00 | Coffee Break & Posters | |
| 3.45 | Poultry Behavior & Welfare 2 | Dairy Behavior & Welfare 2 |
| 4.30 | Closing of Conference | |
| 5.00 | Finish | |
| 6.00 | Farewell Party at Victory Field, Indianapolis | |

# Theater presentations – titles and presenting author

## Monday 1st August

| | |
|---|---|
| 8.00-9.30 | Registration Open |
| 9.00-9.30 | Opening Ceremony |

| | |
|---|---|
| **9.30-10.00** | **Plenary 1:** |
| | **Pain, Distress & Humane End-Points** |

Sex differences in lamb pain sensitivity develop after birth
– Ngaio Beausoleil

| | |
|---|---|
| 10.00-10.45 | **Coffee Break & Posters (odd numbers)** |

| | |
|---|---|
| **10.45-12.00** | **Parallel Sessions** |

| | **Session 1:** | **Session 2:** |
|---|---|---|
| | Pain, Distress & Humane End-Points 1 | Companion Animal Behavior & Welfare |
| 10.45 | Effects of age on piglet distress associated with euthanasia by carbon dioxide or by a carbon dioxide:argon mixture<br>– Larry Sadler | To wee or not to wee: hospitalised female canines (*Canis familiaris*) preferred Astroturf to concrete in a two-way simultaneous presentation choice test<br>– Louise Buckley |
| 11.00 | I'm not going there! using conditioned place preference to assess the aversiveness of restraint and blood sampling in piglets<br>– Puja Wahi | Do you think I ate it? Behavioral assessment and owner perceptions of 'guilty' behavior in dogs<br>– Julie Hecht |
| 11.15 | Non-steroidal anti-inflammatory drugs to mitigate pain in lame sows<br>– Kathleen Tapper | Assessing quality of life in kennelled dogs<br>– Jenna Kiddie |
| 11.30 | Validation of assessment of nociceptive responses in pigs' skin<br>– Pierpaolo Di Giminiani | The development of a behavior assessment to identify 'amicable' dogs<br>– Tammie King |
| 11.45 | 2011 Update of the AVMA guidelines on euthanasia<br>– Gail Golab | Unsocialized or simply scared? The validity of methods commonly used to determine socialization status of shelter cats at intake<br>– Katherine Miller |

| | |
|---|---|
| 12.00-1.30 | **Lunch** |

| | |
|---|---|
| **1.30- 2.00** | **Plenary 2:** |
| | **Maternal Behavior & Effects** |

Echoes from the past: does maternal heat stress adjust offspring to high temperature? An experiment in quails
– Rie Henriksen

| 2.00-3.00 | **Parallel Sessions** | |
|---|---|---|
| | **Session 1 (continued):** | **Session 3:** |
| | Pain, Distress & Humane End-Points 2 | Maternal Behavior & Effects |
| 2.00 | Quantitative sensory testing for assessing wound pain in livestock<br>– Sabrina Lomax | Parturition progress and behaviours in dairy cows with calving difficulty<br>– Alice Barrier |
| 2.15 | Exposure to negative events induces chronic stress and increases emotional reactivity in sheep<br>– Alexandra Destrez | Vocal communication in cattle (*Bos taurus*): mother-offspring recognition<br>– Monica Padilla-De La Torre |
| 2.30 | Castration as a model for studying pain-triggered behavioral responses in growing calves<br>– Lily Edwards | Reproductive and physiologic effects of bedding substrate on ICR and C57BL/6J mice<br>– Melissa Swan |
| 2.45 | Castration as a model for studying pain-triggered cardiac response in growing calves<br>– Luciana Bergamasco | Maternal care and selection for low mortality affect immune competence of laying hens<br>– Bas Rodenburg |
| 3.00-3.45 | **Coffee Break & Posters (even numbers)** | |
| 3.45-5.00 | **Parallel Sessions** | |
| | **Session 1 (continued):** | **Session 4:** |
| | Pain, Distress & Humane End-Points 3 | Communication & Cognition |
| 3.45 | Effect of time, parity and meloxicam (Metacam®) treatment on general activity in dairy cattle during the puerperal period<br>– Eva Mainau | Are auditory cues useful when training American mink (*Neovison vison*)?<br>– Pernille Svendsen |
| 4.00 | The effect of a topical anaesthetic wound dressing on the behavioural responses of calves to dehorning<br>– Crystal Espinoza | A development of a test to assess cognitive bias in pigs<br>– Antonio Velarde |
| 4.15 | The relationship between fearfulness at a young age and stress responses in the later life of laying hens<br>– Elske De Haas | Learning how to eat like a pig: effectiveness of mechanisms for vertical social learning in piglets<br>– Marije Oostindjer |
| 4.30 | A case study to investigate how behaviour in donkeys changes through progression of disease<br>– Gabriela Olmos | Affective qualities of the bark vocalizations of domestic juvenile pigs<br>– Winnie Chan |
| 4.45 | A novel approach of pain recognition and assessment in donkeys: initial results<br>– Neville Gregory | What do ears positions tell us about horse welfare?<br>– Fureix, Carole |

# Tuesday 2nd August

| 8.30-10.15 | **Parallel Sessions** | |
|---|---|---|
| | **Session 5:** Engineering Environments & Measurement Technologies for Science and Welfare | **Session 6:** Qualitative Behavioral Assessment |
| 8.30 | Environment and the development of feather pecking in a commercial turkey facility – Stephanie Torrey | Inter-observer reliability of Qualitative Behaviour Assessment on farm level in farmed foxes – Leena Ahola |
| 8.45 | The effect of cage design on mortality of white leghorn hens: an epidemiological study – Joe Garner | Qualitative Behavioural Assessment can detect artificial manipulation of emotional state in growing pigs – Kenny Rutherford |
| 9.00 | Remedies for the high incidence of broken eggs in furnished cages: effectiveness of increasing nest attractiveness and lowering perch height – Frank Tuyttens | The welfare of pigs in five different production systems in France and Spain: assessment of behavior – Deborah Temple |
| 9.15 | Non-cage laying hen resource use is not reduced by wearing a wireless sensor after habituation – Courtney Daigle | The inter-observer reliability of qualitative behavioural assessments of sheep – Clare Phythian |
| | | **Session 7:** Poultry Behavior & Welfare 1 |
| 9.30 | Effect of exit alley blocking and back-up incidences on the accessibility of an automatic milking system – Jackie Jacobs | Gregarious nesting as a response to risk of nest predation in laying hens – Anja Riber |
| 9.45 | Change-of-state dataloggers were a valid method for recording the feeding behavior of dairy cows using a Calan Broadbent Feeding System – Peter Krawczel | Astroturf® as a dustbathing substrate for laying hens – Gina Alvino |
| 10.00 | An acclimation and handling protocol for implementation of GPS collars for monitoring beef cattle grazing behavior – Angela Green | Feather pecking and serotonin: 'the chicken or the egg?' – Marjolein Kops |
| 10.15-11.00 | **Coffee Break & Posters (odd numbers)** | |
| 11.00-12.00 | **Wood-Gush Memorial Lecture** | |
| | Social Behavior: An Emergent and Adaptive Property of the Mammalian Autonomic Nervous System – Stephen Porges | |
| 12.00-6.00 | **Lunch & Excursions** | |

## Wednesday 3rd August

|  | | |
|------|-----|-----|
| **9.00-9.30** | **Plenary 3:** | |
| | **Laboratory Animal Behavior, Welfare & Enrichment** | |
| | Modifications induced by an enriched environment on reproductive physiology and postnatal development of Albino Swiss mice<br>– Marina Ponzio | |
| **9.30-10.15** | **Parallel Sessions** | |
| | **Session 8:**<br>Laboratory Animal Behavior, Welfare & Enrichment 1 | **Session 9:**<br>Feeding Behavior & Welfare 1 |
| 9.30 | Effects of predictability on feeding and aversive events in captive rhesus macaques (*Macaca mulatta*)<br>– Daniel Gottlieb | Tail biting alters feeding behavior of victim pigs<br>– Elina Viitasaari |
| 9.45 | Exercise pens as an environmental enrichment for laboratory rabbits<br>– Lena Lidfors | The effects of diet ingredients on gastric ulceration and stereotypies in gestating sows<br>– Stephanie Wisdom |
| 10.00 | Does the presence of a human affect the preference of enrichment items in young isolated pigs?<br>– Shelly Deboer | Energy balance and feeding motivation of sheep in a demand test<br>– Amanda Doughty |
| **10.15-11.00** | **Coffee Break & Posters (even numbers)** | |
| **11.00-12.00** | **Parallel Sessions** | |
| | **Session 8 (continued):**<br>Laboratory Animal Behavior, Welfare & Enrichment 2 | **Session 9 (continued):**<br>Feeding Behavior & Welfare 2 |
| 11.00 | Playful handling before an intra-peritoneal injection induces a positive affective state in laboratory rats<br>– Sylvie Cloutier | Effect of milk feeding level on the development of feeding behaviour patterns in dairy calves<br>– Emily Miller-Cushon |
| 11.15 | The naked truth: breeding performance in outbred and inbred strains of nude mice with and without nesting material<br>– Christina Winnicker | Behavioural patterns of dairy heifers fed different diets<br>– Angela Greter |
| 11.30 | Unpredictable repeated negative stimulations modulates effects of environmental enrichment in birds<br>– Agathe Laurence | The effect of lameness on feeding behavior of dairy cows<br>– Petro Tamminen |
| 11.45 | Effects of conditioning on blood draw in cats<br>– Jessica Lockhart | How do different amounts of solid feed in the diet affect time spent performing abnormal oral behaviours in veal calves?<br>– Laura Webb |
| **12.00-1.30** | **Lunch** | |

| 1.30-2.00 | **Plenary 4:** | |
|---|---|---|
| | **Social Behavior in Swine** | |
| | Oxytocin reduces separation distress in piglets when given intranasally – Jean-Loup Rault | |
| **2.00-3.00** | **Parallel Sessions** | |
| | **Session 8 (continued):** | **Session 10:** |
| | Laboratory Animal Behavior, Welfare & Enrichment 3 | Sow Housing |
| 2.00 | Is hair and feather pulling a disease of oxidative stress? – Giovana Vieira | Policy changes to enable sows to express behavioural needs in intensive housing conditions – Cheryl O'Connor |
| 2.15 | Implications for animal welfare: habituation profiles of 129S2, 129P2 and 129X1 mouse strains – Hetty Boleij | Determining the floor space requirement for group housed sows – Fiona Rioja-Lang |
| 2.30 | Identification methods in newborn C57BL/6 mice: a developmental and behavioural evaluation – Magda Joao Castelhano-Carlos | Alleyway width in a free-access stall system influences gestating sow behavior and welfare – Laurie Mack |
| 2.45 | Behavioral and physiological thermoregulation in mice with nesting material – Brianna Gaskill | Action-reaction: using Markov analysis to elucidate social behavior when unacquainted sows are mixed – Jeremy Marchant-Forde |
| **3.00-3.30** | **Coffee Break & Posters (all)** | |
| 3.30-5.30 | ISAE Annual General Meeting | |
| **7.00-11.00** | **Congress Banquet at Hyatt Regency Hotel** | |

## Thursday 4<sup>th</sup> August

| | |
|---|---|
| **9.00-9.30** | ***Plenary 5:*** **Zoo Animal Behavior, Welfare & Enrichment** |
| | Behavior in natural and captive environments compared to assess and enhance welfare of zoo animals<br>– Paul Koene |

| **9.30-10.15** | **Parallel Sessions** | |
|---|---|---|
| | ***Session 11:***<br>Zoo Animal Behavior, Welfare & Enrichment 1 | ***Session 12:***<br>Dairy Behavior & Welfare 1 |
| 9.30 | Effects of enclosure size and complexity on captive African elephant activity patterns<br>– Nancy Scott | Separating the stressors: a pilot study investigating the effect of pre-mixing calves on the behavior and performance of dairy calves in a novel environment<br>– Amy Stanton |
| 9.45 | Gender differences in stereotypical behavior can be predicted by gender differences in activity in okapi<br>– Deborah Fripp | Dairy welfare in three housings systems in the upper Midwest<br>– Marcia Endres |
| 10.00 | Personality and stereotypy components in okapi<br>– Jason Watters | The effect of distance to pasture on dairy cow preference to be indoors or at pasture<br>– Gemma Charlton |

| **10.15-11.00** | **Coffee Break & Posters (all)** | |
|---|---|---|
| **11.00-12.00** | **Parallel Sessions** | |
| | ***Session 11 (continued):***<br>Zoo Animal Behavior, Welfare & Enrichment 2 | ***Session 13:***<br>Human-Animal Interactions |
| 11.00 | Individuals interacting with environmental enrichment: a theoretical approach<br>– Becca Franks | Owner visitation: Clinical effects on dogs hospitalized in an intensive care unit<br>– Rebecca Johnson |
| 11.15 | Do impoverished environments induce boredom or apathy in mink?<br>– Rebecca Meagher | Robot milking does not seem to affect whether or not cows feel secure among humans<br>– Sine Norlander Andreasen |
| 11.30 | Effects of shade on feeding behaviour and feed intake of female goat kids<br>– Lorenzo Alvarez | Relationship between amount of human contact and fear of humans in turkeys<br>– Naomi Botheras |
| 11.45 | Zoo-housed chimpanzees and gorillas are highly selective in their space use: Implications for enclosure design, captive management and animal welfare.<br>– Stephen Ross | Characteristics of stockperson interactions with pigs in swine finishing barns<br>– Sara Crawford |

| **12.00-1.30** | **Lunch** |
|---|---|

| 1.30-2.00 | **Plenary 6:** | |
|---|---|---|
| | **Engineering environments & measurement technologies for science and welfare** | |
| | Perceptual threshold for cold stress in dairy cows | |
| | – Lindsay Matthews | |
| **2.00-3.00** | **Parallel Sessions** | |
| | **Session 11 (continued):** | **Session 14:** |
| | Zoo Animal Behavior, Welfare & Enrichment 3 | Free Papers |
| 2.00 | The effect of diet on undesirable behaviors in zoo gorillas<br>– Elena Hoellein Less | Animal abuse and cruelty: an evolutionary perspective<br>– Emily Patterson-Kane |
| 2.15 | How animals win the genetic lottery: biasing birth sex ratio results in more grandchildren<br>– Collette Thogerson | Behavioural and physiological methods to evaluate fatigue in sheep following treadmill exercise<br>– Michael Cockram |
| 2.30 | Reliability and validity of a subjective measure to record changes in animal behaviour over time<br>– Joanna Bishop | Do differences in the motivation for and utilisation of environmental enrichment determine how effective it is at eliminating stereotypic behaviour in American mink?<br>– Jamie Dallaire |
| 2.45 | The positive reinforcement training effect: reduction of an animal's latency to respond to keepers' cues<br>– Samantha Ward | The effect of pasture availability on the preference of cattle for feedlot or pasture environments<br>– Caroline Lee |
| **3.00-3.45** | **Coffee Break & Posters (all)** | |
| **3.45-4.30** | **Parallel Sessions** | |
| | **Session 15:** | **Session 16:** |
| | Poultry Behavior & Welfare 2 | Dairy Behavior & Welfare 2 |
| 3.45 | Mobile laying hens<br>– Sabine Gebhardt-Henrich | Physiological and behavioral response of crossbred zebu dairy cows submitted to different shade availability on tropical pasture<br>– Carlos Machado Filho |
| 4.00 | Open water provision for pekin ducks to increase natural behaviour requires an integrated approach<br>– Marko Ruis | Changes is dairy cattle behaviour as a result of therapeutic hoof block application<br>– Janet Higginson |
| 4.15 | Does water resource type affect the behaviour of pekin ducks (*Anas platyrhynchos*)?<br>– Donald Broom | Effect of different environmental conditions in loose housing system on claw health in Finnish dairy cattle<br>– Johanna Haggman |
| 4.30-5.00 | Closing of Conference | |
| 6.00-10.30 | Farewell Party at Victory Field, Indianapolis | |

# Abstracts

## Session 02. Companion animal behaviour and welfare

## Session 03. Maternal behaviour and effects

## Session 04. Communication and cognition

## Session 05. Engineering environments & measurement technologies for science and welfare

## Session 06. Qualitative behavioural assessment

## Session 07. Poultry behaviour and welfare 1

## Session 08. Laboratory animal behavior, welfare & enrichment

## Session 09. Feeding behaviour and welfare

## Session 10. Sow housing

## Session 11. Zoo animal behavior, welfare & enrichment

## Session 12. Dairy behaviour and welfare 1

## Session 13. Human-animal interactions

## Session 14. Free papers

## Session 15. Poultry behaviour and welfare 2

## Session 16. Dairy behaviour and welfare 2

## Session 17. Poster session

# David Wood-Gush Memorial Lecture

## Social behavior: an emergent and adaptive property of the mammalian autonomic nervous system

*Porges, Stephen W., University of Illinois at Chicago, Department of Psychiatry, 1601 W. Taylor Street, Chicago, IL 60612, USA; sporges@uic.edu*

The presentation will introduce the Polyvagal Theory as a new perspective that relates the autonomic function to behavior. This approach elaborates on the identification of neural circuits involved in the regulation of autonomic state and an interpretation of autonomic reactivity as adaptive within the context of the phylogeny of the vertebrate autonomic nervous system. The theory enables the investigation of new questions, paradigms, and explanations regarding the role that autonomic function has in the regulation of adaptive physiological states and social behavior. Foremost, the theory emphasizes the importance of phylogenetic changes in the neural structures regulating the heart and how these phylogenetic shifts provide insights into the adaptive function of both physiology and behavior. The theory emphasizes the phylogenetic emergence of two vagal systems: a potentially lethal ancient circuit involved in defensive strategies of immobilization (e.g. death feigning) and a newer mammalian circuit linking the heart to the striated muscles of the face that is involved in both social engagement behaviors and in dampening reactivity of the sympathetic nervous system and the HPA-axis.

## Sex differences in lamb pain sensitivity develop after birth

*Guesgen, Mirjam[1], Beausoleil, Ngaio[1], Minot, Ed[1], Stewart, Mairi[2] and Stafford, Kevin[1],*
*[1]Massey University, Private Bag 11-222, Palmerston North 4442, New Zealand, [2]AgResearch*
*Limited, Animal Behaviour and Welfare, East Street, Private Bag 3123, Hamilton, New Zealand;*
*N.J.Beausoleil@massey.ac.nz*

The sensitivity of the mammalian nervous system to noxious inputs is known to differ according to sex. In addition, there is some evidence that pain sensitivity varies with postnatal age, at least in altricial species such as rodents and humans. In accordance with this, older lambs exhibit higher frequencies of pain-related behaviour after castration/docking than do younger lambs, suggesting that they are more sensitive to pain. However, such differences may also reflect differing abilities to express pain-related behaviour or variation in tissue damage created by these procedures at different ages. This is the first study to explore age and sex effects on the baseline pain sensitivity of sheep. We tested pain sensitivity of 75 lambs aged between 1 and 12 days old (38 males, 37 females); each lamb was tested at only one age. Thermal nociceptive thresholds were measured three times over one hour using a laser device. The beam was aimed at a shaved patch of skin above the coronary band of a hind limb until a withdrawal response was elicited (or 15 s elapsed). The effect of sex on average logged 'latency to respond' was analyzed with age as a linear covariate (ANCOVA). While no overall effects of lamb sex or age were found, there was a significant sex x age interaction ($F_{1,74}=7.2$, P=0.009): pain sensitivity of males and females diverged with increasing age. Older females were more sensitive to pain than were younger females (shorter latencies; r=-0.37 P=0.02). In contrast, the sensitivity of male lambs tended to decrease with age (longer latencies; r=0.25 P=0.14). Postnatal development of sex differences in pain sensitivity may relate to the gradual removal of placental neuroinhibitors. Our findings may have implications for interpretation of previous age-related differences in pain behaviour of lambs and for animal welfare recommendations relating to painful husbandry procedures for sheep.

**Echoes from the past: does maternal heat stress adjust offspring to high temperature? An experiment in quails**

*Henriksen, Rie[1,2], Groothuis, Ton[1] and Rettenbacher, Sophie[2], [1]University of Groningen, Nijenborgh 7, 9747 Groningen, Netherlands, [2]University of Veterinary Medicine, Veterinaerplatz 1, 1210 Vienna, Austria; riehe@sol.dk*

Whereas maternal stress has often been reported to reduce the phenotypic quality of the offspring it has also been suggested that these maternal effects prepare offspring for a stressful environment (Gluckman and Hanson, 2004). Heat stress is a worldwide problem in poultry production. Low body weight gain, reduced egg quality and increased daily mortality are some of its effects. The magnitude of the problems varies between farms and individual birds also react differently to heat stress. While there is no doubt that farm structure and the way farm staff handles heat stress are important factors, part of this variation may be due to maternal effects, a topic hardly addressed in commercial farming. We wanted to test whether maternal heat stress impairs or prepares offspring for a hot environment. We heat stressed (HS=33-35 °C, 7 hours/day) 20 female Japanese quails for 3 weeks, and kept 20 other female Japanese quails at control temperature (C=22 °C, 24 hours/day). Eggs were collected from both groups of females and hatched in the same incubators. Half of the offspring from each groups of mothers were heat stressed (HS) themselves whereas the other half were housed at control temperature (C). Multilevel analyses (MlwiN 1.10.0007) was used for statistics. HS- females laid smaller eggs (P=0.023) and the chicks that hatched from these eggs were smaller (P=0.010) the first 2 weeks of life. From 2 weeks of age, HS-offspring were smaller (P=0.028) independent of mother's treatment. HS-offspring of HS-mothers had a higher respiratory quotient than HS-offspring of C-mothers (P=0.0135) indicating a difference in the utilization of energy resources. Offspring of HS-mothers drank more in the early morning (P=0.049) independent of their own treatment which indicates prenatal behavioural programming for the anticipation of a hot day. HS-offspring of HS-mothers had lower corticosterone response (P=0.046) to an ACTH challenge than HS-offspring of C-mothers, indicating a lower response to a sudden stressor. Whether these finding suggest an adaptive effect of maternal stress will be discussed together with findings on the offspring's immunocompetence.

## Modifications induced by an enriched environment on reproductive physiology and postnatal development of Albino Swiss mice

*Ponzio, Marina F, Luque, Eugenia, Ruiz, Ruben D, Fiol De Cuneo, Marta and Martini, Ana Carolina, Physiology Institute, Medicine Faculty, National University of Córdoba, Córdoba, Argentina, Santa Rosa 1085, X5000ESU, Argentina; ponziomarina@hotmail.com*

Although a growing body of evidence indicates that environmental enrichment (EE) facilitates normal development and behaviour in laboratory mice, few studies have been conducted to demonstrate its impact upon male and female reproductive physiology. In the present study we investigated the effects of PVC tubular devices and shredded paper as physical enrichment on reproductive physiology and postnatal development of laboratory mice. Animals were allocated in regular housing cages in groups of five individuals, and treated as non-enriched (control, C) or enriched from weaning to adulthood (E). In males, parameters evaluated were body, testicular and accessory glands weight, sperm quality (motility, viability, acrosome and membrane integrity), testosterone concentration, *in vivo* fertilization rates and litter size. In females, parameters assessed included body, uterine and ovary weight, spontaneous ovulation, estradiol concentration, pregnancy percentages and litter size. Also, at postnatal day 1 litter was reduced to 8 pups (4 males and 4 females) and their neurobiological (cliff avoidance, negative geotaxis, surface righting reflex), physical (body weight evolution, bilateral pinna detachment, low incisor eruption and eyes opening) and reproductive development (testicular descent, balano-prepucial separation and vaginal opening) was assessed. A second group of C females was enriched from day 1 of pregnancy to discriminate the mother's ability to breed their offspring (EP). A higher number of pups were born from enriched mothers (Mean±SEM; C: 9.5±0.6, n=4; E: 10.7±0.2, n=5; EP: 12.2±0.7, n=5; P=0.03 E and EP vs. C). As well, a strong tendency was detected towards a faster physical and reproductive development of pups born from E and EP mothers, yet significant differences were only observed for testicular descent (day 19, C: 0±0%, n=16; E: 62.5±12.5%, n=17; EP: 21.6±9.7%, n=21; P=0.002). In conclusion, in this strain of mice an increased environmental complexity showed limited effects upon reproductive physiology and postnatal development.

## Oxytocin reduces separation distress in piglets when given intranasally

*Rault, Jean-Loup[1,2], Carter, Sue[3], Garner, Joseph[1], Marchant-Forde, Jeremy[2], Richert, Brian[1] and Lay, Don[2], [1]Purdue Univ., 125 S Russell, 47907 W Lafayette IN, USA, [2]USDA-ARS-LBRU, 125 S Russell, 47907 W Lafayette IN, USA, [3]Univ. of Illinois at Chicago, 1601 W Taylor, 60612 Chicago IL, USA; jrault@purdue.edu*

Oxytocin (OT) is one of the neurobiological foundations of sociality. It acts as a neuropeptide in numerous social processes, from ultra-social to anti-social behaviors. Evidence supports a role for OT in social support, possibly by attenuating separation distress. Nonetheless, research on OT is lacking in our effort to understand social behaviors of domestic animals and implement practices that meet their social and psychological needs. Separation from conspecifics is particularly stressful to social animals, especially prior to weaning. Here, we tested the hypothesis that OT administered intranasally could reduce social separation distress of suckling piglets. At 13 days of age, across 6 litters, 12 piglets received 0.25 ml (24 IU or 50 µg) of OT intranasally (OT) while 12 littermates received 0.25 ml of saline (SAL), balanced by gender. At 45 min after treatment, each piglet was fitted with a telemetric heart rate belt and placed in an isolation box for 15 min. Behaviors, vocalizations and heart rate were recorded during the test and blood was sampled 24 hr before, right after, and 30 min after the test to measure cortisol concentrations. Results were analyzed using a mixed model in SAS.Oxytocin piglets displayed reduced locomotor activity (P=0.02), explored less (P=0.03), spent more time lying (P=0.02) and inactive (P=0.03) than SAL piglets. They also tended to emit fewer grunts and escape attempts (both P<0.1) compared to SAL piglets. Some responses were sexually dimorphic (P<0.05 to 0.1), with females responding stronger than males to OT, possibly due to an estrogen influence. Cortisol concentrations and heart rate did not differ between treatments (P>0.1). Our results support the hypothesis that OT given intranasally can modify behavior, possibly through direct effects on the nervous system. This is the first evidence that OT is able to attenuate social separation distress in farm animals. It is possible that OT is naturally released in the presence of conspecifics, resulting in effects similar to those seen in this study.

## Behavior in natural and captive environments compared to assess and enhance welfare of zoo animals

*Koene, Paul, Wageningen UR, Wageningen UR Livestock Research, Marijkeweg 40, 6709 PG Wageningen, Netherlands; paul.koene@wur.nl*

Wild animals are adapted to the environment they evolved in. In relatively stable environments competition between and within species can make animals specialists (food, defense, etc.); in variable environments animals have to be more adaptive (generalists). Species with specific environmental adaptations may show specific behavioral needs and difficulty in adapting to the captive environment. Animals in zoos are perceived as representatives of their wild counterparts. The discrepancy between natural behavior needs and behavioral possibilities in captivity may be the cause of their welfare problems. In the Netherlands the relation between zoo animal behavior and zoo environment is studied for many years. The aim of this study is to assess behavior and welfare, suggest environmental changes and find species characteristics that underlie zoo animal welfare problems. First, the status of zoo animal welfare assessment is reviewed and the current approach is outlined. Databases of literature on species' natural behavior (1) and captive behavior (2) have been made. Species' characteristics are grouped in eight functional behavioral ecological fitness-related categories (criteria), i.e. space, time, metabolic, safety, reproductive, comfort, social and information needs (subdivided in 68 sub criteria). Assessments of the strength of behavioral needs in relation to environmental demands are made. The literature databases are coupled with databases of behavioral observations (3) and welfare assessments (4). Behavioral data from MSc projects covering 10 Dutch zoos and 45 species are in the observation database. Results of behavioral comparisons of animals in different zoos show, for example, that stereotypies in tigers are related to enclosure size, in giraffes to amount of browse provided and in wolves to the distance to visitors. Welfare assessment methods are based on findings of the Welfare Quality® project. Currently data on 25 most common animal species in Dutch zoos (mammals, birds and reptiles) are collected and the results will be presented. In conclusion, the comparison of the complete behavioral repertoire of behaviors in natural and captive environments promises to highlight behavior and welfare problems, the solution of welfare problems (environmental change) and the species characteristics involved in the causation of the welfare problems.

**Perceptual threshold for cold stress in dairy cows**

*Matthews, Lindsay [1] and Bryant, Jeremy[2], [1]AgResearch, Animal Behaviour and Welfare, Private Bag 3123, 3204 Hamilton, New Zealand, [2]Farmax Ltd, Farm Systems, P.O. Box 1036, 3204 Hamilton, New Zealand; lindsay.matthews@agresearch.co.nz*

This study aimed to determine the perceptual threshold for cold stress in dairy cattle using a trade-off procedure in which shelter could be accessed by giving up a highly-valued activity (resting). Varying levels of cold challenge were arranged by exposing wet animals (n=18) to different wind speeds at winter temperatures. At a minimum of weekly intervals, animals were exposed to up to six cold treatments for 48 h. During the cold challenge, cows were also exposed to either of two levels of lying deprivation (0 or 24 h). Following each cold exposure, wet cows were individually given a mutually-exclusive choice between Resting (roofed, windy enclosure) and Shelter (roofed, calm enclosure, no rest possible) for 60 min. Measures taken included ambient conditions (temperature, wind speed, rainfall), energy intake, animal body condition, weight and hair depth, and time spent in each choice area (and a transition zone). The level of cold challenge was calculated from from the USA National Research Council (NRC) cold stress model and grouped into three bands (Cold – lower than the Lower Critical Temperature (LCT), Mod – within 0-5 degrees of LCT, Warm – >5 degrees above LCT). The time allocated to resting or sheltering in the choice test was analysed as a function of the degree of cold stress with ANOVA. The proportion of time spent sheltering was significantly higher ($P<0.01$) under Cold conditions and did not differ between Mod and Warm (Cold – 0.34, Mod – 0.13, Warm – 0.07); there was no interaction with rest-deprivation treatment. Under rest-deprived treatments, the proportion of time spent resting was significantly higher under Warm conditions (0.41) than Cold (0.17) ($P<0.01$). As rest deprivation is a highly-valued activity to dairy cattle, these results indicate that sheltering becomes critical (from a cow's point of view) under conditions that correspond to temperatures below the LCT.

## Effects of age on piglet distress associated with euthanasia by carbon dioxide or by a carbon dioxide:argon mixture

*Sadler, Larry[1], Hagen, Chad[2], Wang, Chong[3], Widowski, Tina[4] and Millman, Suzanne[1,3], [1]Iowa State University, Biomedical Sciences, 1600 S 16th Street, 50011 Ames, IA, USA, [2]Value-Added-Science & Technologies, 3782 9th Street, SW, 50401, USA, [3]Iowa State University, Veterinary Diagnostic & Production Animal Medicine, 1600 S 2nd Street, 50011 Ames, IA, USA, [4]University of Guelph, Animal & Poultry Science, 50 Stone Road E, N1G 2W1 Guelph, Ontario, Canada; ljsadler@iastate.edu*

The objective of this study was to compare the effectiveness of gases administered for euthanasia between two age groups of piglets: neonates (less than 3 days, n=160, BW 2.61±0.81 kg) and weaned (16 to 24 days, n=160, BW 4.62±0.76 kg). Two different gases were explored in this study: 100% $CO_2$ and a 50:50 $CO_2$:Argon (CA) gas mixture. Each gas was administered at three different flow rates: chamber volume exchange rates per minute of 35%, 50% and Prefill + 20%. A control treatment administered ambient air followed by blunt force trauma. Male-female piglet pairs were placed in a chamber with lid and one side made of clear plastic to facilitate behavior observations. A Smartbox device (Euthanex Corp, Palmer, PA) was used to supply gas at controlled rates. Latencies for behavior and physiologic changes were observed directly, including loss of posture (LP), last movement (LM), gasping (GASP), open mouth breathing (OMB), defecation (DEF), oral behavior (OB) and nasal discharge (ND). Analyses of data were performed in R (v2.12.0, The R Foundation for Statistical Computing) as the univariate product-limit estimation of the survival curves, to determine significant differences. Values are given as raw means and percentages. Neonate piglets were euthanized as quickly as or faster than weaned piglets for all gases and flow rates (latency (sec): LP 99 vs 142 (P=0.001); LM 360 vs 392 (P=0.045); GASP 97 vs 139 (P<0.001); for neonate and weaned piglets, respectively). Main effect of age was observed for the proportion of piglets displaying distress or discomfort for two of the four measured behaviors (% displaying: DEF 23 vs 46 (P<0.001); ND 4 vs 14% (P=0.017); OMB: 97 vs 94 (0.116); OB 25 vs 40 (P=0.116); for neonate and weaned piglets, respectively). Gas by age interactions were observed. Differences were observed between the two age groups, with neonates succumbing to the gas effects faster than weaned piglets.

**I'm not going there! using conditioned place preference to assess the aversiveness of restraint and blood sampling in piglets**

*Wahi, Puja, Widowski, Tina and Yue Cottee, Stephanie, University of Guelph, 50 Stone Road East, Guelph, Ontario, N1G2W1, Canada; pwahi@uoguelph.ca*

Conditioned place preference (CPP) is used to determine the degree of positive/negative reward associated with specific experiences by pairing the experiences with distinct locations and measuring animals' avoidance of or attraction to the locations. 12 weaned piglets were used in each of 3 experiments to investigate the relative aversiveness of restraint and blood collection using CPP. Exp 1 examined whether piglets developed a preference for a pen they explored with a littermate (E) over one where they were restrained in a v-restrainer (R). Exp 2 tested preferences for pens in which piglets were either restrained (R) or restrained and blood sampled from the suborbital sinus (SO). Exp 3 tested preferences for pens in which piglets were restrained and blood sampled from either the suborbital sinus (SO) or jugular vein (JV). The two test pens had distinct floor types and high contrast wall markings. Paired t-tests were used to compare the duration of time that piglets spent in each pen during 4 min pre and post-conditioning tests following 5 days of paired conditioning. In Exp 1 piglets developed a clear preference for pen E (Pre-conditioning: 113.2+14.0 sec in E vs. 126.8+14.0 sec in R; P=0.640; Post-conditioning: 180.9+9.0 sec in E vs. 59.1+9.0 sec in R; P<0.01). In Expt 2 times spent in SO and R pens during both pre and post-conditioning tests did not differ (Pre-conditioning: 140.5+25.6 sec in SO vs. 99.5+25.6 sec in R; P=0.441; Post-conditioning: 107.1+28.3 sec in SO vs. 132.9+28.3 sec in R; P=0.657) but piglets tended to spend less time during the post-conditioning test in pen SO compared to the pre-conditioning test (Pre-conditioning: 140.5+25.6 sec vs. Post-conditioning: 107.1+28.3 sec; P=0.079). Exp 3 showed no preference for either pens (Pre-conditioning: 117.5+19.6 sec in SO vs. 122.5+19.6 sec in JV; P=0.90; Post-conditioning: 103.8+14.9 sec in SO vs. 136.2+14.9 sec in JV; P=0.30). These data indicate that CPP can be used to determine the relative preference/aversion for different handling experiences (Exp 1); that piglets distinguish between restraint only and restraint with blood collection (Exp 2) but do not find blood sampling from the SO any more or less aversive than sampling from the JV (Exp 3).

## Non-steroidal anti-inflammatory drugs to mitigate pain in lame sows

*Tapper, Kathleen[1], Johnson, Anna[1], Karriker, Locke[1], Stalder, Kenneth[1], Coetzee, Johann[2], Parsons, Rebecca[1] and Millman, Suzanne[1], [1]Iowa State University, 1600 S. 16th St., Ames, IA 50011, USA, [2]Kansas State University, 1800 Denison Ave., Manhattan, KS 66506, USA; smillman@iastate.edu*

Objectives were to evaluate effectiveness of sodium salicylate (SS) and flunixin meglumine (BanamineÒ) (FM) to mitigate pain associated with transiently induced lameness in sows. Lameness was induced using Amphotericin B, a model with known consistency, mechanism, intensity and duration of pain. Sows were weight bearing at all times, and lameness resolved within 7 days. Twelve mixed parity crossbred sows received each of 3 analgesic treatments: 1. SS (35 mg/kg q.12.h + 0.04 ml/kg IM sterile saline for handling consistency), 2. FM (2.2 mg/kg IM q.24.h), or 3. Control (C; 0.04 ml/kg IM q.24.h sterile saline). Each sow received each treatment over 3 trials, with 14 day washout periods between trials. Sows were anesthetized and injected with amphotericin B in medial and lateral interdigital spaces on one hindleg. Forty-eight hours post-induction, treatments were administered daily for 4 days. Pain was assessed using pressure algometry (PA), which quantified mechanical nociceptive thresholds (MNT) as kilograms of force (kgf) relative to a foot-lift response. Triplicate measures were taken at 3 landmarks on each hindlimb. Thermal sensitivity (TS) at the coronary band measured latency for a foot-lift response. The observer was blind to pain test output. Data were collected on Day+1 and Day+6 post-induction. Proc Glimmix in SAS 9.2 was used to analyze differences between sound and lame legs. On Day+1, sows displayed 3× greater PA pain sensitivity on lame leg vs. sound leg regardless of treatment (Raw Mean±SEM kgf: Lame 2.1±0.17; Sound 7.7±0.17, P=0.0023). No differences were observed for simple effect comparisons on Day+6 between FM vs. C (P=0.9001), FM vs. SS (P=0.1707), or SS vs. C (P=0.0667). The lame leg increased MNT from Day+1 to Day+6 for all treatments, showing some natural resolution (Day+6 Lame Raw Means±SEM kgf: FM 5.4±0.34; SS 4.6±0.37; C 5.3±0.41). Decreased latency for TS occurred on Day+1 on both sound and lame legs, and was therefore not a valid assessment tool. In conclusion, neither analgesic treatment reduced pain for sows with induced lameness as tested with pressure algometry.

**Validation of assessment of nociceptive responses in pigs' skin**

*Di Giminiani, Pierpaolo[1], Herskin, Mette S.[1] and Petersen, Lars J.[2], [1]Aarhus University, Animal Health and Bioscience, Blichers Alle 20, 8830 Tjele, Denmark, [2]Viborg Hospital, Clinical Physiology, Heiberg Alle 4, 8800 Viborg, Denmark; pierpaolo.digiminiani@agrsci.dk*

Pain perception in pigs can be expressed by several types of behaviours. However, there is still a need to develop new methodologies targeting specific features of nociception. Our current project targets general methodological aspects of pain measurement in normal and inflamed porcine skin through the application of thermal ($CO_2$-laser) and mechanical (Pressure Application Measurement) pain sensitivity devices. The initial stage of the study aimed at comparing measures of pain sensitivity in normal skin. We compared two distinct experimental groups: one comprising 24 pigs weighing $30\pm5$ kg (small) and one of 24 pigs weighing $60\pm5$ kg (big). Within each individual, measurements of sensitivity were taken at two anatomical locations (flank vs. hind legs). Pain sensitivity was compared by recording the latency to respond to the laser and the mechanical pressure at which the animals would show a clear behavioural response. Each experimental animal belonging to the two groups received 4 thermal and 4 mechanical stimulations on each body site and a median response value was calculated. The median latency to respond to the laser was significantly lower in the small animals compared to the big ones at both anatomical sites: 5 (2 to 7) s vs. 12 (6 to 17) s in the flank and 3 (3 to 4) s vs. 6 (3 to 9) s in the hind legs ($P<0.05$). The median mechanical sensitivity in the flank tended to be higher in small compared to big pigs 790 (443 to 1145) g vs. 1195 (510 to 1487) g, ($P=0.06$). Mechanical sensitivity in the legs was significantly lower in the small pigs 287 (206 to 387) g compared to the big ones 444 (314 to 740) g ($P<0.05$). This study is the first to report data from Pressure Application measurements in pigs, and provides valuable indications for selecting the most appropriate body size and anatomical location. It will enable the subsequent phase of the study with the assessment of nociceptive responses in porcine skin following cutaneous inflammation.

**2011 Update of the AVMA guidelines on euthanasia**

*Golab, Gail and Patterson-Kane, Emily, American Veterinary Medical Association, Animal Welfare Division, 1931 North Meacham Rd, Suite 100, 60173 Schaumburg, Illinois, USA; ggolab@avma.org*

Since 1963 the AVMA has convened a Panel on Euthanasia to evaluate methods and potential methods of euthanasia for the purpose of creating guidelines for veterinarians who carry out or oversee the euthanasia of animals. More than 70 individuals, including veterinarians and non-veterinarians with expertise across a range of disciplines and species, were engaged to research and create the 2011 update to the Panel's report (its eighth edition) titled the 'AVMA Guidelines on Euthanasia.' Euthanasia techniques should result in rapid loss of consciousness followed by cardiac or respiratory arrest and the ultimate loss of brain function. In evaluating methods of euthanasia, the Panel used the following criteria: (1) ability to induce loss of consciousness and death with a minimum of pain distress, anxiety or apprehension; (2) time required to induce loss of consciousness; (3) reliability; (4) safety of personnel; (5) irreversibility; (6) compatibility with requirement and purpose; (7) emotional effect on observers or operators; (8) compatibility with subsequent evaluation, examination, or use of tissue; (9) drug availability and human abuse potential; (10) compatibility with species, age, and health status; (11) ability to maintain equipment in proper working order; (12) safety for predators/scavengers should the carcass be consumed; (13) legal requirements; and (14) environmental impacts of methods or carcass disposition. The various sections of the Guidelines address particular euthanasia techniques (e.g., inhalant agents, non-inhalant pharmaceutical agents, and physical methods) and the application of those techniques to various animal types, species, and uses (e.g., companion animals, food animals, laboratory animals, wildlife, aquatics). This edition of the Guidelines has been expanded and includes more detail about the techniques, covers more species, and more comprehensively considers the special needs and challenges posed by the range of environments and conditions under which euthanasia is conducted. This presentation will focus on areas in which the Guidelines have been revised or expanded, the conceptual basis for these modifications, and their anticipated consequences in practice. Attention will be drawn to the need for ethological research to resolve areas of uncertainty and ambiguity.

## Quantitative sensory testing for assessing wound pain in livestock

*Lomax, Sabrina, Espinoza, Crystal and Windsor, Peter, University of Sydney, Faculty of Veterinary Science, PMB 3, Camden 2570, Australia; sabrina.lomax@sydney.edu.au*

Quantitative sensory testing (QST) is a widely used, validated technique that we have adapted for our research to record the evolution and distribution of pain from mulesing, tail docking and/or castration wounds and their response to topical anaesthetic (TA). It is an objective, repeatable form of pain assessment enabling the assessor to distinguish between various analgesic interventions. Von Frey monofilaments are calibrated to bend at a predetermined pressure in order to provide repeatable pain stimulation of predetermined sites on the wound and surrounding skin. Responses are scored using a customized numerical rating scale (NRS), by monitoring induced involuntary motor reflexes in the rump and head. An electronic anaesthesiometer was used to measure pain by means of pressure transduction. Pain threshold was automatically recorded as maximum pressure (g) exerted before animal response and withdrawal. We performed QST using mechanical stimulation with both von Frey monofilaments and electronic anaesthesiometer to evaluate hypersensitivity after surgery in lambs (castration, tail docking, mulesing) and calves (castration). A strong decrease in mechanical thresholds proximal and distal to the wound was indicative of pain. Results indicate that significant wound anaesthesia is achieved within 1 min of mulesing and castration. Three trials have shown that lambs treated with TA had significantly lower mean NRS response scores to von Frey stimulation of their mulesing, castration and/or tail-docking wounds up to 24 hours post-surgery (NRS score ?2, $P<0.01$). Untreated lambs had significant development of hypersensitivity within 1 min of surgery (NRS score ?15, $P<0.01$). In a fourth trial, we found that beef calves treated with TA had significantly lower mean NRS scores to castration, and significantly higher mechanical pressure threshold of the wound than untreated calves ($P<0.01$). This presentation will use our findings to explain how quantitative sensory testing provides us with an important tool for understanding the generation and development of wound pain, and for clinically assessing and quantifying animal pain. With this novel method of pain assessment we have shown that TA can be easily incorporated into farm husbandry procedures to significantly improve livestock welfare.

## Exposure to negative events induces chronic stress and increases emotional reactivity in sheep

*Destrez, Alexandra[1], Deiss, Véronique[1], Leterrier, Christine [2], Boivin, Xavier[1] and Boissy, Alain[1], [1]INRA, UR1213 Herbivores, Adaptation et Comportements Sociaux, 63122 Saint-Genes-Champanelle, France, Metropolitan, [2]INRA, UMR 85 Physiologie de la Reproduction et des Comportements, Comportement, Neurobiologie, Adaptation, 37380 Nouzilly, France, Metropolitan; alexandra.destrez@clermont.inra.fr*

In modern farming systems animals may be repeatedly exposed to aversive events. In humans repeated exposure to aversive events may lead to chronic stress by particularly altering emotional reactivity. The present study investigated whether in sheep, repeated exposure to aversive events induces i) chronic stress and ii) altered reactivity to acute emotional stressors. The study used 48 five-months-old female lambs. Over a period of 6 weeks, 24 of them (treatment group) were submitted daily to unpredictable and inescapable aversive events (various predator auditory and odor signals of carnivores and stressed conspecific, negative handling). The other 24 were reared without these events (control group). Leukocyte counts, resting heart rate and basal cortisol concentrations were determined before and after the treatment period. Emotional reactivity was assessed before and after the treatment period, by submitting lambs to 3 tests, exposure to suddenness, novelty, and human presence. Data were analyzed using the SAS PROC MIXED procedure followed by post-hoc comparisons (Least Square Means Differences using Tukey-Kramer adjustment). Before the treatment period, controls and treated lambs did not differ in physiological values or emotional reactivity. After the treatment, compared to controls, treated lambs had a lower leukocyte counts ($8.5\pm0.27$ vs. $9.6\pm0.32$ x 10 cubed per mm, $P=0.01$), heart rates ($106.0\pm3.4$ vs. $125.0\pm3.3$ bpm; $P=0.0005$) and cortisol levels ($7.3\pm1.1$ vs. $10.2\pm1.0$ ng/ml; $P=0.06$), suggesting that a state of chronic stress had been induced. In addition, treated lambs approached less often the human ($P<0.0001$), vocalized more during exposure to novelty ($P<0.001$) and spent less time in area of the test arena where the sudden event had taken place ($P=0.03$). Results show that repeated exposure to unpredictable and inescapable aversive events may induce chronic stress in lamb, and increased emotional reactivity to acute stressors. Future studies will assess whether these two phenomena are related.

**Castration as a model for studying pain-triggered behavioral responses in growing calves**

*Edwards, Lily[1], Coetzee, Johann[2], Bello, Nora[3], Mosher, Ruby[2], Cull, Charley[2] and Bergamasco, Luciana[2], [1]Kansas State University, Department of Animal Sciences, Manhattan, KS 66506, USA, [2]Kansas State University, Department of Clinical Sciences, Manhattan, KS 66506, USA, [3]Kansas State University, Department of Statistics, Manhattan, KS 66506, USA; lne@ksu.edu*

The objective of this study was to compare pain-triggered behavior of calves of different ages after surgical castration. Male Holstein calves were classified into age groups (n=10): <6 wk (1.5M), 3 mon (3M) and 6 mon (6M). Simulated castration (SHAM) was performed on all calves. Twenty-four h after SHAM, calves were surgically castrated (REAL) following standard U.S. industry procedures without the provision of analgesia. Calves were restrained in a chute during and for 20 min post-treatment. Foot stamps, tail flicks, vocalizations, kicks, collapses and eliminations were recorded during the post-treatment periods. Frequencies of behaviors were analyzed using a generalized linear mixed model including fixed effects of age, treatments and their interaction, and random effects to recognize appropriate experimental units. Across age groups, REAL castration caused a decrease in the frequency of tail movements compared to SHAM (LSMean ± SEM: 36±14 & 183±59, respectively; P=0.0007). The frequency of foot stamps following REAL castration decreased compared to SHAM (3±1 & 6±1, respectively, P<0.0017) for all ages. Regardless of treatment, 1.5M calves vocalized more frequently than 6M and 3M calves (3±2, 0±0 & 0±0, respectively; P<0.05). For frequency of kicks, an age*treatment interaction was identified (P=0.02) whereby 3 M and 6M, but not 1.5M, calves showed an estimated decrease of 2 kicks following REAL castration compared to the SHAM. The frequency of collapses post-treatment differed between ages (P=0.001); 1.5M calves collapsed more frequently than the 3M and 6M calves (2±0.3, 1±0.2 & 0±0, respectively; P<0.0016). The probability of elimination was low with no significant differences between ages or treatments (P>0.42). For all ages, we observed a decreasing frequency of active behaviors such as foot stamps and tail flicks during post-castration suggesting a period of reduced activity following a painful stimulus. Other active behaviors were decreased only among older calves, whereas passive behaviors were more frequent in the youngest calves. Results support age-specificity in behavioral responses to acute pain.

## Castration as a model for studying pain-triggered cardiac response in growing calves

*Bergamasco, Luciana[1], Edwards, Lily [2], Bello, Nora[3], Mueting, Stacy [1], Cull, Charley[1], Mosher, Ruby[1] and Coetzee, Hans[1], [1]Kansas State University, 1800 Denison Ave, 66506 Manhattan KS, USA, [2]Kansas State University, 232 Weber Hall, 66506 Manhattan KS, USA, [3]Kansas State University, 101 Dikens Hall, 66506 Manhattan KS, USA; bluciana@vet.k-state.edu*

In animals heart rate variability (HRV) is used to assess cardiac autonomic nervous system regulation and response to painful procedures. Aim- To explore HRV in calves before and after simulated and surgical castration. Animals- Thirty male Holstein calves, 3 age categories (<6 weeks=1.5M; 3 months=3M; 6 months=6M; n=10 calves/group). Procedure- After a 5 d of acclimation, calves were submitted to a simulated-castration session (S) followed 24 h later by surgical castration (R) without any pain medication provided. For each session, heart rate (HR) was continuously monitored. HRV measures: mean HR (beats/min-bpm), square root of the mean squared differences of successiveinter-beat-intervals (RMSSD-high frequencyvariation estimation-), high frequency power (HF-vagal activity-), low frequency power (LF-sympathetic activity-), LF/HF ratio. HRV analysis: times (t) -1 (before procedure) and at +5, +10 and +20 min post-treatment (512 inter-beat-intervals-around 5 min-). Statistical analysis: general linear mixed model including age, treatment, time effects and all interactions. Results- HR: No significant treatment*time*age interaction; significant age*time (P<0.0048) and treatment*time (P<0.0001) interactions. HR decreased at t+5 (99.2 bpm) vs t+20 (106.9 bpm; P=0.012) in 1.5M calves. HR decreased at t+5 (77.14 bpm) vs t+20 (84.3 bpm; P=0.00038) in R, and t+5 decreased in R (77.14 bpm) vs S (92.01 bpm; P=0.012). RMSSD: no significant treatment*time*age and treatment*time interactions; age*time interaction (P=0.02). RMSSD increased at t+10 (61.5 ms) vs t+20 (39.8 ms) in 1.5M calves (P=0.036). HF, LF, LF/HF: no significant interaction between age, time and treatment. Marginal treatment effect (P=0.12) on HF: R showed greater HF (9.06%) vs S (7.24%). Age effect (HF, P=0.0067;LF, P=0.0266;HF/LF, P=0.003): 6M calves showed lower HF (2.17%), higher LF (34.5%), and higher LF/HF (7.09) vs 1.5M (18.6%,8.7%,1.15 respectively) and 3M (13.17%,18.4%,1.6, respectively) calves. Preliminary results show a decline in HR immediately after R across age groups which may be associated to a shift toward vagal activity related to deep visceral pain.

## Effect of time, parity and meloxicam (Metacam®) treatment on general activity in dairy cattle during the puerperal period

*Mainau, Eva, Cuevas, Anna, Ruiz-De-La-Torre, José Luis and Manteca, Xavier, UAB, School of Veterinary Science, 08193 Bellaterra (Barcelona), Spain; eva.mainau@irta.cat*

The objective of this study was to investigate the effect of time, parity and the non-steroidal anti-inflammatory drug meloxicam on general activity around eutocic calving, as possible indicator of pain and discomfort in dairy cattle. Sixty Friesian dairy cows from first to sixth parity with calving that required no or light assistance were included. Cows were randomly allocated into two homogeneous groups regarding parity and treated with either meloxicam (Metacam® 20 mg/ml inj.sol; Boehringer Ingelheim) subcutaneous, at a dose of 0.5 mg/kg of body weight or placebo as control. Treatments were administered within a maximum of 6 hours after calving (mean ± SE: 4.35±0.988 h). General activity around calving was obtained by recording the cows' posture and the number of steps. Postures (standing, lateral and semi-lateral lying, walking and changing position) were observed at 10 minutes intervals using video recordings from 2 days before (day-2) to 2 days after calving (day+2). The number of steps per day was obtained using activity meters (Westfalia®, Germany) from day-1 to day+7. Statistical analyses were performed with SAS using a GENMOD procedure. Both parameters confirmed that cows were more active during the days preceding calving and on the day of calving (day0) than after. Cows spent more time displaying active behaviour (standing, walking and changing posture) during the 2 days before calving and more time displaying passive behaviour (lateral and semi-lateral lying) just after calving ($P<0.0001$). The number of steps was higher from day-1 to day+2 than from day+3 to day+7 ($P<0.05$) and the maximum value occurred on day0 (mean ± SE: 113±9.7 steps/h), with a 40% increase in activity compared with baseline values (80±5.8 steps/h on day+7). A parity per time and a parity per treatment effect interaction were found on number of steps ($P<0.05$). Heifers performed more number of steps than multiparous cows from day-1 to day+2. Moreover, heifers in the meloxicam group showed a significantly higher number of steps during the days around calving than heifers from control group (112±5.3 and 91±3.1 steps/h respectively). The effects of meloxicam on general activity around calving may result from the anti-inflammatory effects of meloxicam and assist in reducing pain associated with calving.

**The effect of a topical anaesthetic wound dressing on the behavioural responses of calves to dehorning**

*Espinoza, Crystal, Windsor, Peter and Lomax, Sabrina, The University of Sydney, Faculty of Veterinary Science, 425 Werombi Road, Camden 2570, Australia; crystal.espinoza@sydney.edu.au*

Dehorning is an important yet painful procedure in the cattle industry routinely performed without pain relief. Topical anaesthesia may be suggested as a novel method of pain relief potentially effective at addressing the economic and practical constraints of current methods. Aim: to assess the effect of a novel topical anaesthetic wound dressing on the behavioural responses of calves to dehorning. Thirty 2-month-old Holstein-Friesian heifer calves were crush-restrained and randomly allocated to sham scoop dehorning (C), scoop dehorning (S), or scoop dehorning with an immediate post-procedural application of a topical anaesthetic (ST) (modified Tri-Solfen®, Bayer Animal Health, Australia) (no cautery). Quantitative sensory testing was performed to assess efficacy of the topical anaesthetic by measuring the extent of sensation and/or anaesthesia of the dehorned wound and adjacent skin area. Testing involved the use of von Frey monofilaments (instruments calibrated to bend at specific pressures) to provide light touch (10 g) and pain stimulation (300 g) and observing responses. Calf head and rump involuntary reflexes and motor responses were graded depending on vigour such that nil, mild, moderate and severe responses (Y) were allocated values 0, 1, 2 and 3 respectively. Testing was performed before, and at 1 and 40 min, 1.5, 4 and 24 h post treatment Data were analysed using ordinal logistic regression to evaluate the effect of treatment on response severity. Dehorned and treated (ST) calves were 4 (Prob=0.001) and 3 (Prob=0.002) times less likely to display severe responses (Y=3) than S calves 40 min (Prob=0.005) and 1.5 h (Prob=0.007) post treatment (P=0.002 and 0.02 respectively). Tendency for ST calves to display less severe responses than S calves 1 min and 24 h post treatment. Dehorned calves (S) were more likely to display more severe responses (Y=1-3) than C calves at all time points post treatment (P<0.05). Similar responses were found between C and ST calves 1 and 40 min post treatment, changing thereafter with ST calves displaying more severe responses (P<0.05). Topical anaesthetic was able to reduce sensitivity of the dehorned wound site at some time points up to 24 h post scoop dehorning in calves. Intervention addressing procedural pain is needed.

**The relationship between fearfulness at a young age and stress responses in the later life of laying hens**

*De Haas, Elske N[1], Kops, Marjolein S[2], Bolhuis, Elizabeth J[1] and Rodenburg, Bas T[3], [1]Wageningen University and Research Centre, Adaptation Physiology Group, Marijkeweg 40, 6700 PG Wageningen, Netherlands, [2]Utrecht University, Department of Pharmaceutical science / Division Pharmacology Utrecht University, Sorbonnelaan 16, 3584 CA Utrecht, Netherlands, [3]Wageningen University and Research Centre, Animal Breeding and Genomics Centre, Marijkeweg 40, 6700PG Wageningen, Netherlands; elske.dehaas@wur.nl*

Stress responses of birds can be affected by their fear level. In young chicks inactivity in an Open Field (OF) is indicative of fearfulness, while corticosterone elevations after restraining adult hens are often used to assess response to an acute stressor. We were interested whether chicks' OF behavior is related to their response to restraining when adult. A low mortality line (selected on low mortality due to feather pecking (FP)) and an unselected control line were used: eight pens of 10 birds each. Chicks were individually subjected to a 10-min OF test at 7 wk. of age, and to 5-min restraining (MR) at 35 wk. of age. Directly after the MR test, blood samples were collected to assess plasma corticosterone levels. Effects of line were analyzed with pen as experimental unit, using a GLM-procedure. Activity levels during MR test, which could elevate corticosterone levels, were corrected for. Pearson correlations between the residuals (corrected for line effects) of OF and MR variables were calculated at pen level. Chicks from the low mortality line were more active in the OF than chicks from the control line (P<0.01). Lines did not differ in behavior during the MR test or in corticosterone response to MR. Pens that were active in the OF at 7 wk. of age had a reduced corticosterone response at 35 wk. of age (at pen level r=-0.71, P<0.01). As said, low activity of young birds in the OF is indicative of fearfulness. Activity of young chicks was – on pen level – negatively correlated with corticosterone after restraining. These findings may indicate a relationship between fearfulness in young chicks and stress responses as adults. This in turn may be affected by the group they live in. The difference seen in the OF behavior of different genetic lines may imply a reduction of associated factors that induce feather pecking, and may help in understanding the causal relationship between fear and FP.

## A case study to investigate how behaviour in donkeys changes through progression of disease

*Olmos, Gabriela[1,2], Mc Donald, Gemma Adele[1], Elphick, Florence[1], Neville, Gregory[1] and Burden, Faith[2], [1]Royal Veterinary College, Hawkshead Lane, Hatfield, Hertfordshire, AL9 7TA, United Kingdom, [2]The Donkey Sanctuary, Sidmouth, Devon, EX10 0NU, United Kingdom; goantillon@rvc.ac.uk*

Donkeys have a limited repertoire of non-specific signs displayed when in pain or sick. This study looked closely at donkey behaviour during the progression of different diseases with the aim of improving pain and sickness recognition. Video footage of a group of 79 donkeys at the Donkey Sanctuary was obtained for 6 months; where 45 diseased cases observed. Due data completes, four cases were selected [Cases A) with respiratory disease due to herpes virus (n=2) and Cases B) end-stage cases (hyperlipaemia, n=1; chronic laminitis, n=1)] plus four healthy controls (n=4). Cases A were observed for 8 hrs on day-10 and -1 prior to disease onset (day 0 = first veterinary visit) and during treatment (day 1, 5 and 10). Cases B were observed for 8 hrs on day -7, -3 and on the day of euthanasia (day 0). Total time (minutes) performing 47 different behaviours were compared between (painful/sick vs. healthy) and within donkeys using chi-square or fisher's exacts tests. Diseased donkeys in cases A and B spent on average 10% more time (range, 3-17%, P<0.01) with a lowered head carriage compared to controls. Conversely, they spent 15% less time (range 6-34%, P<0.04) with their ears in combinations (i.e. each ear in opposite direction), thus meaning ears were more static and unresponsive. Ear changes were subtle but were the earliest indicators of pain/sickness in the observed donkeys. Cases B compared to the controls spent 31% more time in recumbency (range 7-60%, P<0.01), and 40% less time eating (range 1-64%, P<0.01). The reduction in total eating time was not substituted by any other oral behaviour (e.g. drinking, grooming, licking, and investigative behaviours), where drinking and grooming were greatly affected in the donkey with hyperlipaemia. Finally, abdominal effort was only observed in cases A and tended to reduce with time on treatment (P=0.06). Donkeys are working animals of great importance worldwide, and these results highlight useful behavioural changes that can be used as monitoring signs of pain/sickness in these animals. The potential use of these signs warrants further studies in greater and more diverse donkey populations.

**A novel approach of pain recognition and assessment in donkeys: initial results**

*Olmos, Gabriela[1,2], Alvarado-Arellano, Ayin Q[3], Dutoit, Nicole[2], Burden, Faith[2] and Gregory, Neville[1], [1]Royal Veterinary College, Hatfield, Hertfordshire, AL9 7TA, United Kingdom, [2]The Donkey Sanctuary, Sidmouth, Devon, EX10 0NU, United Kingdom, [3]UNAM, Depto de Patología – FMVZ, Ciudad Universitaria, DF, 04510, Mexico*

This paper proposes an approach to use pain-relevant pathologies to enhance our understanding of the clinical and behavioural signs of pain in donkeys and outlines initial results of this ongoing investigation. The methodology is summarized as follows. Trained veterinary clinicians examined live donkeys under two situations: A) before being euthanized due to a terminal illness or reduced quality of life (n= 347 sedentary donkeys in UK; DU) or B) when about to be slaughtered in an abattoir (n=164 working donkeys in Mexico; DM). The animals that represent populations in terms of age (years) for DU and DM respectively, were: <5 = 0.5%, 31.7%; 5-15 = 4%; 67.7%; 16-20 = 6%, 0.6%; >20 = 89.5% and 0%. For sex were: stallions 0.5%, 44%; geldings 52.2%, 8%; females 47.3%, 48% for DU and DM respectively. The body condition was: <2 =18.5%, 44.7%; 2.5-3 = 56.5%, 54%; >3.5 = 25%, 1% for DU and DM respectively and the girth was: 115 cm ±SD 9.6, 112 cm ±SD 7.3 for DU and DM respectively. The clinical examination (CE) included the oral mucosa, heart & respiratory rate, rectal temperature, plus an evaluation of 6 demeanours and 47 behaviours/signs that could relate to pain. At this point an overall pain visual analogue score was derived (VAS 0 cm = no pain to 10 cm = the worst pain). At post-mortem (PM), lesions/pathologies were noted, grouped by system-organ/tissue, ranked (mild, moderate, severe) and classified according to the following potentially painful pathologies: (1) trauma, (2) inflammation, (3) over-distension (4) perforation/ rupture, (5) stripping/ulceration, (6) adhesions, (7) swelling, (8) exposure of sub-chondral bone. From these observations a second VAS was produced. Raw correlations from the two populations showed that donkeys given a higher VAS at CE and PM presented with a greater severity of lesions in more systems as well as a higher heart rate at CE than those donkeys with a lower PM VAS. Moderate to severe pain identified CE was often recognised as severe pain at the PM stage. These initial observations show promise, and so further analysis will be done to test the relationships between pain indicators and pain pathologies.

**To wee or not to wee: hospitalised female canines (*Canis familiaris*) preferred Astroturf to concrete in a two-way simultaneous presentation choice test**

*Teer, Sally and Buckley, Louise, Harper Adams University College, Animals Department, Newport, Shropshire., TF10 8NB, United Kingdom; 00701530@harper-adams.ac.uk*

Dogs develop urination substrate preferences and are often reluctant to toilet when confined to a small space. However, many veterinary facilities provide only a concrete-floored toileting area. Many hospitalised dogs appear averse to urinating on concrete and will delay urination or soil their cage. Good patient care involves stressor identification and subsequent removal or management. Therefore, facilities that encourage hospitalised dogs to urinate more readily should be identified. This study investigated female dog preferences for urinating on either fake grass (Astroturf) or concrete. It was predicted that Astroturf would be preferred. 57 bitches were recruited on an ad hoc basis (all hospitalised female dogs during a 3 week period). Dogs were given on-leash toileting opportunities throughout the day in a concrete-floored outside exercise run. Half the run was covered with Astroturf and half was left uncovered. The run side the Astroturf was situated on was switched daily and the run disinfected between urination episodes. Whether the dog urinated, the latency to urinate and the substrate urinated on was recorded. 30 out of 57 dogs urinated at least once during hospitalisation. The mean (± std. dev) latency to urinate on the first occasion was 46.5 seconds (±30.9). (Astroturf was the preferred substrate: 29 / 30 dogs selected Astroturf for their first urination bout (One sample binomial test, $P<0.001$, 95% C.I. 0.8278-0.9992). It is concluded that Astroturf shows promise as an alternative substrate for urination. However, this preference needs additional investigation before fake grass is recommended as an environmental modification. Further research will identify the potential implications of this preference for the welfare and management of hospitalised canines.

## Do you think I ate it? Behavioral assessment and owner perceptions of 'guilty' behavior in dogs

*Hecht, Julie[1] and Gácsi, Márta[2], [1]Royal (Dick) School of Veterinary Studies, Easter Bush, Roslin EH25 9RG, United Kingdom, [2]Eötvös Loránd, Pázmány P. s. 1/c, H-1117 Budapest, Hungary; julblue@gmail.com*

Dog owners ascribe guilt to dogs. Our study investigated an owner-reported anecdote: upon an owner returning home, a dog greets the owner and sometimes displays 'guilty' behavior, thereby alerting the owner to the dog's misdeed. The questionnaire examined owners' perceptions of dog 'guilt'. The experiment explored (1) whether dogs that were disobedient in owners' absences show associated behaviors of 'guilt' (ABs) upon owners' return to a room and (2) whether owners can determine dog disobedience based on dog greeting behavior. Owners (N=64) reported: dogs display ABs (87.5%); ABs imply dogs know that they transgressed (91%) and dog presentation of ABs could lead owners to scold dogs less (59%). The experiment used pet dogs (N=58) and established the social rule that food on a table was for humans. Dogs had the opportunity to eat after the humans left the room. Owners returned, were unable to see the table and therefore observed dog greeting behavior to decide if the dog ate. Behavior analysis revealed no difference in display of ABs during greeting between obedient and disobedient dogs (Mann-Whitney U-test; P>0.050). While owners appeared able to determine whether or not dogs ate in their absence ($\chi^2$=11.266, DF=1, P<0.001), a subset of owners who were most likely to base their assessment on actual greeting behavior, rather than in-test experimental cues, were not better than chance in their determination (Fisher's exact probability test; P>0.050). Their inability to identify obedience by relying on greeting behavior corresponds with our behavior analysis. Regarding the owners' appeared ability to identify obedience or transgression, their experience with dog in-test social-rule ascription or a personalized or holistic approach to assessing behavior could have contributed to their success. Presentation of ABs could serve an adaptive function in dog-human social conflicts during domestication. The attribution of guilt warrants exploration as it could affect interspecific social exchanges. The questionnaire supported the present anecdote, but the experiment did not detect a propensity of dogs to display post-transgression ABs.

## Assessing quality of life in kennelled dogs

*Kiddie, Jenna[1], Mills, Daniel[2], Hayes, William [2], Neville, Rachel[2], Morton, David [3], Pfeiffer, Dirk[1] and Collins, Lisa[4], [1]Royal Veterinary College, London, AL9 7TA, United Kingdom, [2]University of Lincoln, Lincoln, LN2 2LG, United Kingdom, [3]University of Birmingham, Birmingham, B15 2TT, United Kingdom, [4]Queens University Belfast, Belfast, BT7 9AL, United Kingdom; jkiddie@rvc.ac.uk*

The aim of this project was to investigate the quality of life (QoL) of kennelled dogs in rehoming centres with different previous kenneling experience. 48 dogs from two rehoming centres were each tested on 11 days, over a 34 day period. On each sampling day, urine and saliva were collected. Urine was assayed for oxidative stress, cortisol, 5-HT, and selected catecholamine metabolites (all calibrated against creatinine). Saliva was assayed for total antioxidant capacity. Physical condition of dogs was assessed using body, eye, and nose scores and core body and facial temperatures. Video cameras were placed in the dogs' kennels on each sampling day for 2 hours: activity budgets were used to calculate behavioural diversity; and sequential dependency of behaviour. Behaviours were analysed using mean total proportions of time spent in sight. Spearman's rank correlation, Mann-Whitney, Kruskal-Wallis and Chi-square tests were used to analyse variables for between-measure correlations and associations. Correlations with the dogs' age, gender and source (transferred from another centre; first time relinquished; repeat relinquished) were also analysed. The alpha level was set at $P=0.01$ to correct for multiple test effects. Dogs that were transferred from another kennel were easier to handle; those relinquished for the first time avoided handling ($P=0.003$, $\chi^2=13.264$, $N=31$). Transferred dogs tended to eat all of their food; dogs relinquished for the first time varied in the amount they ate ($P=0.012$, $\chi^2=12.823$, $N=28$). There was a trend for first time relinquished dogs to spend longer walking than transferred or returned dogs ($P=0.031$, $H=5.287$, $N=18$). There was a trend for transferred dogs to have higher oxidative stress than first time relinquished dogs ($P=0.082$, $H=4.611$, $N=19$). No significant within-dog changes were found over the study period (Friedman's Chi-square). The results suggest that dogs from different sources may need to be treated differently on entering rehoming centres. Past experience is the most likely reason for this. The results from this study will be used to develop an objective, quantified and validated system for scoring QoL in kennelled dogs.

## The development of a behavior assessment to identify 'amicable' dogs

*King, Tammie[1], Marston, Linda[1] and Bennett, Pauleen[2], [1]Anthrozoology Research Group, Animal Welfare Science Centre, Monash University, Wellington Rd, Clayton, 3800, Australia, [2]School of Psychological Science, LaTrobe University, Edwards Rd, Bendigo, 3552, Australia; tammie.king@monash.edu*

Modern day dogs are kept primarily as companions; few are utilized for working roles. However, many breeds traditionally bred for work still exist and are kept as pets. Inappropriate dog-owner matching may lead to dogs developing behavioral problems that can endanger the public and cause community disruption. Many of these dogs are relinquished to shelters where a large proportion is euthanized. Questionnaire data indicates that most people want 'amicable' dogs. The ability to measure this characteristic could be useful in assessing animals for breeding purposes or rehoming. The Monash Canine Amicability Assessment (MCAA) was developed using a modified version of the Ainsworth's Strange Situation Test, during which the dog is exposed to an unfamiliar environment and person in the presence and then absence of the dog's owner. The protocol was applied to 200 pet dogs. Each dog's behavior was video recorded and behavioral variables (locomotion, orientation, location, respiration, posture, vocalization, tail wagging and contact with human) were scored. High amicability ratings as determined by owner questionnaire results and as rated by two dog behaviorists based on the video footage, were associated with the dog spending less time near the owner's chair in the presence of the stranger (r=- 0.29, n=97, P<0.005), more time near the stranger (r=0.25, n=97, P<0.05) and more time in contact with the stranger when the owner was absent (r=0.25, n=97, P<0.05). Stranger Fear was associated with less contact (r=- 0.27, n=100, P<0.01) and less tail wagging (r=- 0.25, n=100, P<0.05) with the stranger when interaction with the dog was attempted, as well as low body posture (r=0.40, n=100, P<0.001) throughout the assessment. While these are preliminary results, they suggest that the MCAA may provide a more objective way to measure amicability than currently available which could have many applications. The ability to measure aspects of behavior that contribute positively to the ownership experience could enable breeders to selectively breed animals most suited for life in the modern world. In conjunction with educating the public about dog behavior, this may improve the human-dog relationship and the welfare of pet dogs.

**Unsocialized or simply scared? The validity of methods commonly used to determine socialization status of shelter cats at intake**

*Miller, Katherine, Slater, Margaret, Weiss, Emily, Mirontschuk, Alex and Makolinski, Kathleen, American Society for the Prevention of Cruelty to Animals (ASPCA), Community Outreach, 520 Eighth Avenue, 7th Floor, New York, NY 10018, USA; katherinem@aspca.org*

Free-roaming cats are often brought to U.S. animal shelters by people who know very little about them. The handling of a free-roaming cat in a shelter, and the disposition options available for it, depend heavily on where the cat appears to fall along the socialization spectrum from truly unsocialized to well-socialized with humans. However, accurately determining socialization status of a cat entering a shelter can be difficult because many cats behave fearfully upon arrival in the novel shelter environment with unfamiliar handlers. There are currently no validated methods of determining cats' socialization status upon shelter intake. By studying 253 cats whose socialization status was known through a survey of their caregivers, we investigated 57 behavioral and five physical measures taken during 17 assessments for their ability to accurately determine socialization status within 3 days of arrival in a shelter setting. Results indicate that cats' behavior changes over the first 72 hours after arrival altering the ability to accurately predict socialization. Logistic regression modeling indicated that some measures commonly used by animal welfare professionals do not accurately differentiate cats' socialization status within 72 hours of shelter intake. During certain assessments, however, cats' ear position, head movement, body position, vocalization, aggression, eye contact with the observer, and food ingestion could be used to differentiate truly unsocialized cats from semi- and well-socialized cats with at least 75% accuracy.

**Parturition progress and behaviours in dairy cows with calving difficulty**

*Barrier, Alice, Haskell, Marie and Dwyer, Cathy, SAC, Easter Bush, Midlothian EH25 9RG, United Kingdom; alice.barrier@sac.ac.uk*

The welfare of dairy cows and their calves is compromised following a difficult calving. A better understanding of what happens during a difficult calving is needed to help prevent and alleviate adverse health and production effects. The objective of this study was to investigate parturition behaviours and calving progress of cows, either giving birth normally or with difficulty. Video footage of calvings leading to singleton liveborn calves over 15 months were considered for the study. 12 FN (farm assisted no calf malpresentation) selected at random and the 7 available FM (farm assisted with calf malpresentation) were each paired to a non-assisted calving with respect to dam's parity, sex and birth weight of the calf, genetic line, and calving season. The 38 calvings were observed on three distinct continuous periods relative to full expulsion of the calf: -6 h to 5 h 30 (A); -4 h to -3 h (B) and -2 h to birth (C). Labour lengths (duration from appearance of calves feet until birth) did not differ between scores of difficulty (median time in min; N: 54.7; FN: 101.3; FM: 194.0; P>0.05) but there was large individual variability. As early as period B, FN and FM cows displayed more contractions than N cows and this was also the case for FN cows in period C but not for FM cows (P<0.05). FN cows were also more restless (counts of postural transitions) during periods B and C (+65% and + 33% respectively; P<0.05) despite increased levels of restlessness regardless of calving difficulty (P<0.05). Overall, FM cows raised their tail for longer (in % of observation time; N: 33.7±4.2; FN: 42.7±5.1; FM: 54.0±7.0; P<0.05) compared to N cows, and FN cows tended to lie down for longer durations (P<0.10). There also was a tendency of FM cows to lie lateral with their head rested more than N cows during period B and FN cows to do so in period C (P<0.10). There was no effect of calving difficulty on self-grooming, feeding, lying to standing transitions, exploratory (lick ground and sniffing) or 'irritation' behaviours (stamping, tail switching, rubbing, turning head back). Cows with calving difficulty showed altered parturition progress and expressed some of the behaviours differently over the course of parturition. This may relate to different pain levels when dystocia occurs and could also be used towards early detection of calving difficulty.

## Vocal communication in cattle (*Bos taurus*): mother-offspring recognition

*Padilla-De La Torre, Monica[1], Ochocki, Brad[1], Briefer, Elodie[2], Reader, Tom[1] and Mcelligott, Alan G[2], [1]University of Nottingham, Animal behaviour and Ecology, University Park, Nottingham, NG7 2RD, United Kingdom, [2]Queen Mary University of London, School of Biological and Chemical Sciences, Mile End Road, London, E1 4NS, United Kingdom; plxmp4@nottingham.ac.uk*

It is crucial in order to avoid misdirected maternal investment and to ensure survival of young. In ungulates, parent-offspring identification is often achieved through vocal communication. The most common vocalizations in mother-offspring interactions are contact calls. Mothers and their offspring emit contact calls in order to find each other when they are separated. This type of communication represents a highly important process due to the strong bond between the two partners involved. Among ungulates vocal communication in cattle is still poorly understood. However, a few studies have suggested that calls may reflect the physiological and emotional state, and/or motivations and intentions of the calling animal. Considering then that cattle vocalizations are probably meaningful to other members of the herd, we tested the hypothesis that mother-offspring individual vocal recognition exists. Mutual mother-offspring vocal recognition is expected to be more important in follower species, where offspring follow their mothers rapidly and mingle in groups of unrelated conspecific, than in hider species, where offspring lie concealed in vegetation during the first weeks after birth. The study was carried out with a beef cattle herd living freely on a farm in Nottingham, UK. Mutual mother-offspring vocal recognition was tested in the field using playback experiments (n=22 mother-offspring pairs). We scored three levels of behavioral response to each playback: (1) ears twitching, (2) head turning, (3)walking towards speaker, and (4) vocalizing and/or moving towards the speaker. Our results showed that cows were more likely to respond (at all levels) to calls of their own calves than to calls from unknown calves, ($X^2$>6.70; df=1; P<0.009) but that calves did not show differential responses to their own mothers and unknown cows ($X^2$<2.44; df=1; P>0.117). In the context of the hider/follower dichotomy in ungulates, we suggest that our results are consistent with the theory that hider species display unilateral vocal recognition. Nevertheless, more experiments are needed to establish which anti-predator strategy is/was used by farm cattle and their wild ancestors.

**Reproductive and physiologic effects of bedding substrate on ICR and C57BL/6J mice**
*Swan, Melissa and Hickman, Debra, Indiana Univeristy School of Medicine, Laboratory Animal Resource Center, 917 W. Walnut St IB 008, Indianapolis, IN 46205, USA; mpswan@iupui.edu*

Bedding substrate selection can be a difficult decision for the advisory staff as picking a universal bedding is highly dependent on cost, researcher need, and facility preference. This study examined the effects of bedding substrate on reproductive behaviour and physiology. Thirty six ICR and 36 C57BL/6 mice were placed into monogamous breeding pairs and divided into 3 treatment groups with an n=6 per group. Each group was placed on one of three substrates: Harlan 7097 ¼' corncob bedding, 7093 shredded aspen, or 7084 recycled paper. The pairs were maintained continuously for four months or until the third litter of offspring was weaned. Pups per female per day, a commonly used calculation to monitor the production efficiency of commercial mouse colonies, was compared between bedding substrate using an ANOVA. No statistical significance was found between bedding substrate treatment groups (ICR, P=0.4679; C57BL/6, P=0.3912). No statistical significance was found in the success survivability for each litter, as defined by number of pups weaned divided by number of pups born, between bedding substrate treatment groups when analyzed by ANOVA (ICR, P=0.5999; C57BL/6, P=0.9962). No statistical significance was found in the number of litters produced by each female between bedding substrate treatment groups (ICR, P=0.4490; C57BL/6, P=0.2948) when analyzed by ANOVA. This study demonstrated that there were no statistically significant differences in the reproductive efficiency of these two strains of mice when housed on these three bedding types.

## Maternal care and selection for low mortality affect immune competence of laying hens

*Rodenburg, T. Bas[1], Bolhuis, J. Elizabeth[2], Ellen, Esther D.[1], De Vries Reilingh, Ger[2], Nieuwland, Mike[2], Koopmanschap, Rudie E.[2] and Parmentier, Henk K.[2], [1]Wageningen University, Animal Breeding & Genomics Centre, P.O. Box 338, 6700 AH Wageningen, Netherlands, [2]Wageningen University, Adaptation Physiology Group, P.O. Box 338, 6700 AH Wageningen, Netherlands; bas.rodenburg@wur.nl*

We studied effects of selection on low mortality and of brooding on peripheral serotonergic (5-HT) variables and on immune competence. Previous studies suggest that laying hens selected for low mortality have higher whole-blood 5-HT levels and are better able to cope with fear and stress. Brooding has been shown to have similar positive effects. It is unknown, however, whether this better coping ability also applies to immune challenges. A total of 153 birds was used: half of the birds originated from a line selected for low mortality (L) in group housing, the other half from a control line (C). Within each line, half of the birds was reared with a foster from 0-7 weeks of age (L+ and C+), while the other half was not (L and C). Birds were housed in floor pens in groups of 10. At 56 wk of age, birds were subjected to a 5-min manual restraint test. Fifteen min after the start of the test a blood sample was collected to measure whole-blood 5-HT and platelet 5-HT uptake. At 58 wk of age birds were immunized intratracheally with 0.1 mg human serumalbumine (HuSA). Total antibody titers to HuSA in plasma from all birds were determined by ELISA at days 0, 3, 7, 14, 21 and 31. Data were analysed using ANOVA, testing effects of brooding, line and their interaction and corrected for pen nested within treatment. Antibody titers were analysed using repeated measures ANOVA with polynomial contrasts. Birds from the L line tended to have higher whole-blood 5-HT levels than birds from the C line (P<0.10). Brooded birds tended to have a lower 5-HT uptake levels than non-brooded birds (P<0.10). Within the non-brooded birds, birds from the L line had higher levels of HuSA binding antibodies than birds from the C line (P<0.05), whereas the brooded groups were in between. Furthermore, the C group responded more strongly to immunization than the other groups, followed by C+, L and L+ (quadratic contrast line*brooding; P<0.01). These results indicate that maternal care and selection on low mortality have positive effects on the ability to cope with fear and stress and on immune competence.

**Are auditory cues useful when training American mink (*Neovison vison*)?**

*Svendsen, Pernille, Aarhus University, Faculty of Science and Technology, Dept. of Animal Health and Bioscience, Blichers Allé 20, P.O. Box 50, Dk-8830 Tjele, Denmark; pernille.svendsen@agrsci.dk*

Using auditory cues for animal training present advantages to experimenters as sound can be detected by an animal with less strict orientation towards the cue than otherwise necessary for the detection of visual cues. Mink are naturally active during dark/twilight hours and accordingly have a good auditory sense. In this experiment I investigated how female mink respond to repeated auditory cues, if they can discriminate between them and whether using auditory cues for training is applicable for use in future learning experiments. I tested 15 juvenile female farm mink using an unreinforced habituation-dishabituation technique. Using the technique, a novel cue is presented repeatedly to the animal, to induce habituation and it is then followed by a novel cue to cause dishabituation. Discriminative ability can then be inferred. The test consisted of three successive presentations of sound cue A, followed by sound cue B during the fourth and final presentation. Sound cues were played for 10 s followed by an inter-trial-interval of 60 s, in a balanced design, using tones of 2 and 18 kHz as sound cues. Before testing, the animals were trained until they were confident in the testing apparatus. Behavioural observations, i.e. freezing behaviour and orienting head/body towards sound source, of the reactions towards the sound cues were made. Habituation occurred already with the second sound cue presentation, regardless of tones used ($P<0.01$ for both 2 and 18 kHz as first cue). There were no significant difference in reaction time between sound cue 2 and 3 ($P>0.05$ for both tones). Thus the mink recall and habituate to the sound after one presentation. However, they did not dishabituate to a final novel cue presentation as response time increased on the last cue, regardless of frequency used ($P>0.05$). The mink did not demonstrate discrimination between 2 and 18 kHz. This may, however, be a result of the cue presentations being unreinforced. The findings suggest that using auditory cues for training female mink is applicable. Female mink respond actively to novel sounds and habituate rapidly without any difference between low and high frequency sound cues.

**A development of a test to assess cognitive bias in pigs**

*Mainau, Eva[1], Llonch, Pol[1], Rodríguez, Pedro[1], Catanese, Bernardo[1], Fàbrega, Emma[1], Dalmau, Antoni[1], Manteca, Xavier[2] and Velarde, Antonio[1], [1]IRTA, Finca Camps Armet, 17121 Monells, Spain, [2]UAB, School of Veterinary Science, 08193 Bellaterra, Spain; antonio.velarde@irta.cat*

The aim was to develop a test to assess cognitive bias as an indicator of emotional state in pigs. This study is part of a larger project aimed at assessing the role of cognitive bias in animal emotions and welfare. Thirty-six male pigs (41.47±1.535 kg body weight) were individually trained during 18 sessions to discriminate between a bucket with or without access to chopped apples according to its position (left or right) in a 34 m² test pen. After the training sessions, each animal was subjected individually to an experimental session, where the bucket was placed on a central situation. Both training and experimental sessions finished 30 seconds after the pig ate or tried to eat apples or 10 minutes after the pig entered the pen, which was marked with semicircular lines on the floor from 1 m to 5 m away from the bucket. The time to cross each line, to contact the bucket and to eat or try to eat apples and the number of vocalizations and freezing events (defined as a pig stopped for more than 2 seconds without showing exploratory behaviour) were recorded. Statistical analyses were performed with SAS using a GENMOD procedure. During the training sessions, the time to cross the lines, to contact the bucket and to eat or try to eat apples were significantly lower when the bucket was in the position where access to food was possible (P<0.01). Although the counts of freezing events did not differ significantly between treatments, pigs performed more vocalizations in the sessions without access to food (P<0.01). During the experimental session, 52.78% of pigs were classified as having a positive cognitive bias (time to eat was similar to that during the training sessions with accessible food) and a 16.67% were classified as having a negative cognitive bias (when this time was similar to that without accessible food). A 30.55% of pigs could not be classified. These preliminary results revealed that after training sessions, pigs could predict the availability of food in a bucket depending on its position, where the time to contact the bucket and vocalizations appears as good indicators of predictability. Moreover, this suggests that decision making and behaviour of trained pigs in front of ambiguous situations may be useful to classify them according to its affective state.

**Learning how to eat like a pig: effectiveness of mechanisms for vertical social learning in piglets**

*Oostindjer, Marije[1], Bolhuis, J. Elizabeth[1], Mendl, Mike [2], Held, Suzanne[2], Van Den Brand, Henry[1] and Kemp, Bas[1], [1]Wageningen University, Department of Animal Science, Adaptation Physiology Group, P.O. Box 338, 6700 AH Wageningen, Netherlands, [2]University of Bristol, Department of Clinical Veterinary Science, Centre for Behavioural Biology, Langford House, Langford, BS40 5DU, United Kingdom; marije.oostindjer@wur.nl*

Social learning may help piglets to eat solid food earlier, thereby reducing weaning problems. For applications it is important to know which learning mechanisms are important. Experiment 1 compared learning to eat by observation and by participation. Piglet pairs could observe their mother (O) or participate (P) while she was eating a flavoured feed in a test room for 10 min/day from day 16-20 of age. Pairs of piglets that could eat while the sow was present but not eating (E) and control piglets (C) were only exposed to the test room with the sow were included as treatments. Piglet pairs were tested on days 23-25 for 90 min/day without the sow and could choose between flavoured food eaten by the sow and another flavoured food. Data of pairs were analysed using generalized linear mixed models including treatment, batch and sow flavour as main factors and sow as random factor. O and P piglets had the shortest latencies to eat (O&P: 33±9 min, E&C:60±8 min, P=0.007), a higher consumption on test day 1 (O: 16±3,P: 12±3, E&C: 5±1 g, P=0.001) and higher preference for the flavour eaten by the sow (O&P: 60±4, E&C: 52±4%, P=0.006) than C and E piglets. Experiment 2 compared information about location and food type. Piglets observed the sow eating a flavoured food from one of two feeders on different sides of the room for 10 min/day during five days. In the test phase there was a match or mismatch between location and food type that the sow was eating. Match piglets ate sooner (match: 5.5±1.8, mismatch: 18.7 min ± 6.6, P=0.05) and ate more (match: 37±11, mismatch: 13±4 g, P=0.02) from the feeder where the sow had fed than mismatch piglets. Latency to approach and consumption of sow food type (match: 37±11, mismatch 22±6 g, P=0.6) did not differ between treatments, suggesting that piglets prioritized information of food type over location. Observation, participation and food type are thus important factors for piglets to learn from the sow, and should be considered when designing solutions to reduce weaning problems.

## Affective qualities of the bark vocalizations of domestic juvenile pigs

*Chan, Winnie Y. and Newberry, Ruth C., Washington State University, Center for the Study of Animal Well-being, Department of Animal Sciences and Department of VCAPP, P.O. Box 646351, Pullman, WA, 99164-6351, USA; winnie.chan@email.wsu.edu*

Bark vocalizations are usually emitted by pigs in alarming situations, suggesting that they function as alarm calls. However, juvenile pigs also occasionally produce barks when playing. Although barks given in these two contexts sound similar to the human ear, it is possible that they sound different to pigs, resulting in different behavioural responses. We hypothesized that barks reflect fearful and playful states through differences in acoustic morphology. We analyzed the acoustic structure of barks given by 6-wk-old pigs during alarm and play contexts, and the behavioural responses of randomly-selected focal pigs after the occurrence of these barks. Barks in the alarm context (n=74) were induced by the sudden appearance of an approaching human whereas barks in the play context (n=86) were recorded during spontaneous play. Barks given in playful contexts had lower mean peak frequencies (mean±SE; 0.923±0.047 kHz) than barks given in alarming contexts (1.071±0.042 kHz; ANOVA: $F_{1,158}$=9.90, adjusted P<0.05). Pigs were more likely to look up (0.74±0.16 vs 0±0) and flee (0.73±0.18 vs 0±0), and less likely to scamper (0±0 vs 1.92±0.79), immediately (within 1 s) following barks emitted in the alarm, than the play, context (MANOVA: Hotelling-Lawley Trace = 1.09, $F_{3,26}$=9.43, P<0.05). These data support our hypothesis that differences in the acoustic morphology of barks reflect underlying fearful and playful affective states, and provide evidence that pigs behave differently in response to barks given in these two contexts.

## What do ears positions tell us about horse welfare?

*Fureix, Carole[1], Rochais, Céline[1], Ouvrad, Anne[1], Menguy, Hervé[2], Richard-Yris, Marie-Annick[1] and Hausberger, Martine[1], [1]University of Rennes 1, UMR CNRS 6552 Ethologie Animale & Humaine, Campus de Beaulieu, bâtiment 25, 263 avenue du Général Leclerc, 35042 Rennes cedex, France, [2]Cabinet médical de chiropractie, 1 rue Ernest Psichari, 35136 St Jacques de la Lande, France; carole.fureix@univ-rennes1.fr*

Developing approaches to measure welfare states as objectively as possible, particularly by non-invasive and easily applied methods, remains a scientific challenge. Ear positions, a postural element easily recordable, have been shown to reflect acute stress in several species, and one may wonder if it could also reflect chronic stress. Here we tested the hypothesis that ear position could be a reliable indicator of chronic poor welfare in horses, comparing horses' ear positions in their box to several welfare indicators. Observations were performed on 59 adult horses from riding schools. Ear positions were observed while foraging on the ground only at quiet times in the stables, outside feeding times by a silent experimenter, walking slowly and regularly in the stable, on 2 days at 15 min intervals until 10 scan samples of ear positions were obtained per horse. Chronic welfare state was assessed by chronic health disorders, vertebral problems and behavioural observations of stereotypies. It appeared that 34% of the horses suffered from chronic health disorders, 73% were severely affected by vertebral problems and 66% showed stereotypic activities, evidencing poor welfare states for some of these 59 horses. Interestingly, the 31 horses displaying mostly a backward ear position (i.e. $\geq$ 50% of the 10 scan samples) were prone to suffer from health disorders, to be affected by vertebral problems and to display stereotypies (Fisher, Mann Whitney & Spearman correlations, $P<0.05$). Conversely, the more time spent with forward ears, the less health-related problems and number of stereotypies (Spearman correlations, $P<0.05$). These results both confirm earlier findings in horses, reporting a backward ears position in acute aversive situations and go beyond by showing that a similar backward ears position could be observed more permanently in horses experiencing chronic poor welfare. Conversely, forward ears may also reflect good welfare. At the time where the need has appeared to include both negative and positive emotions in welfare assessment, this study clearly shows that posture may be of great help in both directions.

**Environment and the development of feather pecking in a commercial turkey facility**

*Duggan, Graham[1], Weber, Lloyd[2], Widowski, Tina[1] and Torrey, Stephanie[1,3], [1]University of Guelph, Department of Animal and Poultry Science, Guelph, ON, N1G2W1, Canada, [2]LEL Farms, Guelph, ON, N1L1G3, Canada, [3]Agriculture and Agri-Food Canada, Guelph, ON, N1G2W1, Canada; storrey@uoguelph.ca*

Feather pecking is a serious welfare concern for the poultry industry. While inroads have been made into understanding the causal factors involved in feather pecking in the laying hen, little research has been done to examine the problem in the domestic turkey. Environment appears to play an integral role in feather pecking, although the relationship between environment and pecking has never been examined in commercial housing. The goal of this pilot study was to examine the development of feather pecking in 2 environments in a commercial turkey tom facility. After rearing in identical pens, 49,332 beak-trimmed male turkeys were placed in two growing facilities at 4.5 weeks of age: environmentally controlled (artificial light and ventilation; E) or curtain-sided (natural light and ventilation; C) environment (5000-7500 turkeys/pen; n=8 pens) through 15 weeks. Video-recordings captured feather pecking behaviour, feather condition was scored on 4 body regions (neck, back, wing and tail) and weights were measured on a randomly-selected 100 birds/pen every 3 weeks. Mortalities and culls were recorded as they occurred. Data were analyzed with a mixed-model analysis, with repeated measures where applicable. Mortality and behaviour data were log and square-root transformed, respectively. Light intensity in E ranged from 1-338 Lux. C barns experienced intensities ranging from 150-4800 Lux. We found a difference (P=0.01) in severe feather pecking between the two environments, with 2.8 times as many bouts occurring in C as in E. There was no difference in gentle feather pecking (P=0.84). Feather scores were different between the two environments (P=0.007), with C having worse plumage than E throughout the experiment. Culls and mortality were also influenced by environment (P=0.002). In E, 3.2% of turkeys were culled or died, with 1.1% of culls and deaths due to pecking. In C, 6.5% of turkeys were culled or died, with 4.4% of culls and deaths due to pecking. Growth rates did not differ between environments. In conclusion, the lack of control over the environment in a commercial barn was detrimental to turkey welfare by leading to increased feather pecking and resulting injuries and deaths.

## The effect of cage design on mortality of white leghorn hens: an epidemiological study

*Garner, Joseph P.[1], Kiess, Aaron S.[2], Hester, Patricia Y.[1], Mench, Joy A.[3] and Newberry, Ruth C.[4], [1]Purdue University, 125 South Russell Street, IN 47906, USA, [2]Mississippi State University, Box 9665, MS 39762, USA, [3]UC Davis, One Shields Avenue, CA 95616, USA, [4]Washington State University, 116 Clark Hall, WA 99164, USA; jgarner@purdue.edu*

Many husbandry problems, such as mortality, are particularly difficult to manage because they are influenced by a complex interaction of many factors. Such problems can be intractable to evaluate in a conventional experiment where only one or two factors can be manipulated. Epidemiological approaches provide a potential solution by using multifactorial 'natural' variation between management systems to study husbandry problems. A cross-sectional epidemiological survey combining on-farm measurement and production records was developed and tested on commercial laying hen farms. We then visited a total of 179 houses, of which 167 yielded data suitable for analysis. For each house, mortality was calculated between placement in laying cages and 60 wk of age. For analysis, we prioritized the variables to examine current hypotheses in the field; removed variables without sufficient variation or those with missing data; and identified highly correlated variables and condensed them into single summary variables. We then developed a single hypothesis-led GLM model that best described factors affecting the variance in mortality ($R^2$=62%). Mortality was lower ($P<0.05$): (1) in A-frame than vertical cages; (2) at an optimum floor space of 70 in$^2$ (452 cm$^2$) per hen; (3) in deep versus shallow cages; (4) as feeder space per hen increased; (5) with use of nipple drinkers; (6) in the W36 strain of Leghorn hens; (7) with evaporative cooling; (8) with lower caloric intake; (9) at lower light intensities; and (10) in flocks with cleaner feathers. These results indicate a number of risk factors for mortality associated with cage design as well as genetics, the environment, and diet. They also suggest potential management interventions to reduce mortality for future study.

**Remedies for the high incidence of broken eggs in furnished cages: effectiveness of increasing nest attractiveness and lowering perch height**

Tuyttens, Frank[1], Van Baelen, Marjolein[2], Bosteels, Stephanie[2] and Struelens, Ester[2], [1]Institute for Agricultural and Fisheries Research, Animal Sciences Unit, Scheldeweg 68, 9090 Melle, Belgium, [2]University College Ghent, Brusselsesteenweg 161, 9090 Melle, Belgium; frank.tuyttens@ilvo.vlaanderen.be

Two remedial treatments were investigated to reduce the incidence of broken eggs in furnished cages: increasing the attractiveness of the nest box and lowering the height of the perch (to reduce the chance of egg damage when eggs were laid from the perch). A 2×2 factorial design was used with low (L, 7 cm) or high (H, 24 cm) perches as the first factor, and with nest box floors equipped with either artificial turf (A) or plastic wire mesh (P) as the second factor. Eight cages, each housing on average 8 Lohman Brown hens (aged 40-56 wks), were used per treatment. From 18 wks of age until the start of the experiment, the hens had been familiarized to perches of the other height, but had been exposed only to plastic wire as nesting material. For 61 days the location of egg laying and egg shell cracks were scored. Direct scan-sampling observations of all hens were carried out during 14 days spread out over the experimental period to record hen position (cage floor, nest or perch). In addition, 8 cages (4H + 4L) were videotaped during the light period when hens were 54-56 wks old to record perch use and behaviour. Log-transformed data were analyzed using a two-way ANOVA with cage as the experimental unit. Nesting material influenced the location of the egg cracks (more cracks along the longitudinal side in A vs. P cages, P=0.043), but not the percentage of eggs broken or laid outside the nest. L cages had a lower prevalence of total eggs (P=0.016) and outside nest eggs (P=0.004) broken than H cages. Perch use increased during the observation period, and more so for the H cages during day-time and for the L cages during night-time. Video-analyses revealed that perch bout duration (P<0.001), the likelihood of voluntary ending a perching bout (P=0.013), and the likelihood of sitting (P=0.007) and wing/leg-stretching (P<0.001) were lower in L versus H cages. Lowering perches seems to be a more promising remedy for the high incidence of broken eggs in furnished cages than the provision of artificial turf as a nesting material, but lower perches increase disturbances of day-time perching behaviour.

## Non-cage laying hen resource use is not reduced by wearing a wireless sensor after habituation

*Daigle, Courtney and Siegford, Janice, Mich St Univ, An Sci, 1290 Anthony Hall, E Lansing, MI 48824, USA; lyndcour@msu.edu*

A wireless sensor has been developed to monitor non-cage laying hen space use and activity levels. This sensor was placed inside a casing and mounted on a hen's back with a figure eight nylon harness. The casing was colored to match hen feather color and painted with a unique number for easy visual identification. Hens have complex social structures and body weight, comb size and previous interactions are important in determining social status. However, hens are sensitive to flockmate phenotypic differences, and different looking hens may become feather pecking victims or exhibit different levels of resource use and aggression. Four rooms of 135 hens were weighed and 10 hens/room were randomly selected across body weight distribution and fitted with a sensor at 11 wks of age (d0). Instantaneous scan sampling recorded the number of hens using each resource (feeder, water, nestbox, perch) every 5 minutes over 24 hrs on d-2, d-1, d1, d2, d4, d8, and d16. Tukey-Kramer test determined no difference in overall proportion of hens using the feeder (t=0.19; P=0.853) or water (t=-0.68; P=0.494) after sensor placement. Differences were observed in the proportion of hens using the nestbox (t=-9.12; P<0.0001) and perch (t= -4.75; P<0.0001) after sensor placement. Proportions of casing and non-casing hen resource use were recorded on d1, d2, d4, d8, and d16. Logistic regression determined that feeder use by sensor-wearing hens was less on d1 ($X^2$=12.13; P=0.005) and d2 ($X^2$=6.88; P=0.009), and more on d16 ($X^2$=48.17; P<0.0001) than non-sensor hens. Water use by sensor-wearing hens was reduced only on d1($X^2$=4.80; P=0.029). Nestbox use by sensor-wearing hens increased on d1 ($X^2$=181.64; P=0.0001), d2 ($X^2$=0.65; P=0.0001) and d16($X^2$=75.64; P=0.0001). Sensor-wearing hens perched more on d1 ($X^2$=10.62; P=0.001), d2 ($X^2$=11.01; P=0.001) and d4 ($X^2$=8.97; P=0.003), and less on d8 ($X^2$=20.34; P<0.0001). Initial resource use was affected by wearing a sensor, but by d16 all resources were used similarly or more than by non-sensor wearing hens. No difference in body weight on d16 was observed (t=-0.25; P=0.8) suggesting that long-term resource use was not affected after habituation. Ongoing analysis of agonistic interactions may highlight underlying effects of wearing sensors on resource use and behavior.

**Effect of exit alley blocking and back-up incidences on the accessibility of an automatic milking system**

*Jacobs, Jacquelyn and Siegford, Janice, Michigan State University, Animal Science, 2265 Anthony Hall, East Lansing, MI, 48824, USA; jacob175@msu.edu*

Gates and alleys positioned around an Automatic Milking System (AMS) may impact cow traffic and cow behavior, potentially affecting the system's availability and efficiency. One particular inefficiency problem involves cows in the holding area blocking the exit of cows from the exit alley. Occasionally, this problem escalates from a blocking event, where one cow is prevented from leaving the exit alley, to a back-up event, where two or more cows are prevented from leaving. These back-up events directly influence the availability of the AMS, as the most recently milked cow is unable to exit the milking stall, forcing other cows to wait for access to the AMS. In order to assess the potential reduction in the availability and efficiency of AMS systems due to back-up events, eighty-four lactating Holstein dairy cows were divided into two equal groups balanced for parity and stage of lactation. Each group had access to a single AMS with a gate and alley design that was a mirror image of the other. Cow locations and behaviors in the AMS entrance and exit areas and in the adjacent holding area were recorded continuously for 14 days. Pearson's coefficient of correlation was used to determine the relationship between the duration of successful milking events and potential inefficiencies associated with back-up events. Time spent engaged in back-up events had a weak negative correlation with the duration of successful milking events ($r=0.26$, $P<0.01$). Moreover, time spent engaged in back-up events, unsuccessful milking events, and robot empty events had a strong negative linear correlation with time spent on successful milking events ($r=0.85$, $P<0.001$), with robot empty events accounting for the most variability related to successful milking events ($r=0.94$, $P<0.001$). The AMS was empty an average of 14% of the day, and it was expected that back-up events would be absorbed by the available empty time. However, back-up events and robot empty events had no specific relationship ($r=0.02$, $P=0.42$), suggesting back-up events are not mitigated by the robot being empty and available for milking. This study describes the inefficiencies associated with one particular gate and alley design around the AMS. Information is needed on an ideal gate and alley configuration that will ensure efficient traffic flow through an AMS.

**Change-of-state dataloggers were a valid method for recording the feeding behavior of dairy cows using a Calan Broadbent Feeding System**

*Krawczel, Peter D., Klaiber, Lisa M., Thibeau, Stephanie S. and Dann, Heather M., William H. Miner Agricultural Research Institute, P.O. Box 100, Chazy, NY 12919, USA; krawczel@whminer.com*

Evaluation of feeding behaviors is important for understanding the interaction of diet and welfare for dairy cows, however, its assessment from a Calan Broadbent Feeding System has, historically, required the labor-intensive practices of direct observation or video review. The objective of this study was to validate the output of a HOBO change-of-state datalogger, mounted to the door shell and latch plate, against continuous video data. Data (number of feed bin visits (n/d) and feeding time (min/d)) were recorded using both methods from 26 lactating cows (mean parity = 2.3±0.1; mean days in milk = 20.5±2.5) and 10 non-lactating cows (mean parity = 1.7±0.4; mean days relative to calving = -10.4±2.6) for 3 d per cow (n=108). The agreement of the datalogger and video methods was evaluated using the REG procedure of SAS to compare the mean response of the two methods (video and logger) against the difference between the methods (video minus logger). The maximum allowable difference (MAD) was set at ± 3 for bin visits and ± 20 min for feeding time. From video data, feed bin visits ranged from 2 to 140/d and feeding time from 28 to 267 min/d. Agreement was established between the datalogger and video methods for feed bin visits (P=0.47; $R^2$<0.005), but was not established for feeding time (P<0.001; $R^2$=0.25; y = -0.64x + 92.5) using complete dataset (all data). The combination of a significant P-value and high $R^2$ suggested a slope bias within the data. As a result, this dataset was screened to remove visits of a duration ≤3 sec reflecting a cow unable to enter a feed bin (7% of all data) and ≥5400 sec reflecting a failure of the door to close properly (<1% of all data). Using the resulting screened dataset, agreement was established for feed bin visits (P=0.57; $R^2$<0.003) and feeding time (P=0.13; $R^2$=0.01). For bin visits, 4% of the data was beyond the MAD. For feeding time, 3% of the data was beyond the MAD and 74% of the data was ± 1 min. The insignificant P-value, low $R^2$, and concentration of the data within the MAD validate the usage of a change-of-state datalogger to assess the feeding behavior of cows feeding from a Calan Broadbent Feeding System. Use of the screening criteria for data analysis is recommended.

**An acclimation and handling protocol for implementation of GPS collars for monitoring beef cattle grazing behavior**

*Green, Angela R.[1], Rodriguez, Luis F.[1] and Shike, Daniel W.[2], [1]University of Illinois, Agricultural and Biological Engineering Dept, Urbana, IL, USA, [2]University of Illinois, Dept of Animal Science, Urbana, IL, USA; angelag@illinois.edu*

Implementation of GPS for monitoring beef cattle location has been reported with varying success. Limitations of using this technology have included cost of units, precision, accuracy, sampling frequency, battery life, and maintenance. Previous researchers have developed lower cost GPS collars, but still reported challenges to implementing them. In this work, we: (1) adapted a previous design and constructed low-cost GPS collars (approx. $300 each); (2) developed and assessed a training protocol including acclimation to continuously wearing collars, reward (cracked corn) for cooperative handling, and field approach for daily checks; and (3) characterized individual variations in temperament and handling of each animal in the holding area and the field (1-5 scale, lower = easier to handle). A mixed group of 15 cows and 15 heifers was trained and tested. The training protocol consisted of the cattle wearing, in succession, over 1 week: a collar, a collar with empty box, a collar with weighted box, and, a collar with the GPS unit. Following the training period, cattle were brought into the chute every 5 days to change batteries and download data. For each time the cattle were brought into the chute, they were separated into groups of 10 and allowed to self-sort in the tub before entering. Order through the chute was recorded, and temperament at each phase of the handling process was recorded. In the field, each animal was approached individually and assigned a temperament score for 'approach' and 'touch.' Reward was offered only after allowing a touch on their face, neck, or collar. Over the month-long study, average temperament scores for field approach improved from 4.5±0.5 to 2.1±0.8 and for handling in chute from 2.1±1.3 to 1.0±0.2. Cattle order through the chute could predictably be used to identify the cows that were most difficult to handle. During both tests, all cattle were successfully trained to wear the collars; no GPS units were damaged by the cattle during the study; and 70% of the cattle regularly allowed thorough field inspections. During the final weeks, several collars were adjusted in the field with the addition of wire ties or loosening or removal of collar in the field.

**Inter-observer reliability of Qualitative Behaviour Assessment on farm level in farmed foxes**

*Ahola, Leena Kaarina, Koistinen, Tarja and Mononen, Jaakko, University of Eastern Finland, Department of Biosciences, P.O. Box 1627, FIN-70211 Kuopio, Finland; leena.ahola@uef.fi*

WelFur project aims at developing a Welfare Quality®-like on-farm welfare assessment protocols for farmed fur animals. Preliminary protocol for foxes was developed in 2010. One of the potential measures in the protocol is Qualitative Behaviour Assessment (QBA). The aim of the present study was to analyze inter-observer reliability (IOR) of QBA in farmed foxes on farm level. For QBA, a fixed rating scale of 29 descriptors was designed. An extra descriptor defining each assessor's overall opinion of the welfare of the animals on the farm (OOW) was scored, too. The scales were tested on 18 commercial fox farms in Finland by three assessors with only limited training on QBA scoring but some or extensive experience of farmed foxes. QBA included observing the animals on each farm for 10-15 min, and then scoring the descriptors on farm level using a visual analogue scale. The observed animals on each farm were in most cases not the same individuals for the three assessors. IOR of the QBA scores was analyzed with Kendall Correlation Coefficient W. Despite the small number of assessed farms, Principal Component Analysis (PCA, no rotation) of QBA scores was carried out separately for each assessor. Correlations between principal components (PC1, PC2) and the assessor's OOW were analyzed with Pearson correlation for each assessor. Out of the 29 descriptors, nine (Content, Confident, Happy, Fearful, Immobile, Nervous, Pretentious, Agitated, Passive) got Kendall's W values over 0.55 (for all these descriptors: $P<0.05$). W value for OOW was 0.46 ($P>0.1$). W values for PC1 and PC2 were 0.55 ($P<0.05$) and 0.45 ($P>0.1$), respectively. PC1 was, on average, positively loaded with positive expressions (e.g. Relaxed, Harmonious, Positively occupied, Comfortable) and negatively loaded with negative expressions (e.g. Bored, Fearful, Tense, Nervous). For two assessors, there was a significant correlation ($r=0.59-0.71$, $P<0.01$) between PC1 and OOW. In conclusion, IOR of the present QBA carried out in a true commercial fox farming situation is at most moderate. For two assessors, QBA reflected their view of foxes' overall welfare. The results point out the need of refining the list of QBA descriptors as well as training of assessors on QBA scoring.

**Qualitative Behavioural Assessment can detect artificial manipulation of emotional state in growing pigs**

*Rutherford, Kenny, Donald, Ramona, Lawrence, Alistair and Wemelsfelder, Francoise, SAC, Animal Behaviour and Welfare, Bush Estate, Midlothian, EH26 0PH, United Kingdom; kenny.rutherford@sac.ac.uk*

Scientific assessment of affective state in animals is challenging but vital for animal welfare studies. One possible approach is Qualitative Behavioural Assessment (QBA), which integrates the many perceived aspects of animal demeanour directly in terms of emotional expression (e.g. relaxed, anxious). Previous research has found QBA to have high inter- and intra-observer reliability and good correlation with quantitative measures of behaviour and physiology. Here, we investigated whether QBA could discriminate the effects of a pharmacological manipulation of emotional state. Forty young pigs were treated with Azaperone (a drug with anxiolytic properties in pigs) or saline, prior to either an open field (n=12, balanced cross-over design) or an elevated plus-maze test (saline: n=14, Azaperone: n=14). QBA analysis of two one-minute video clips of each pig was provided by 12 observers, using a Free Choice Profiling methodology, where observers initially generated their own individual list of terms to describe pig emotionality and then scored each recording of pig behaviour for each of these terms (on a visual analogue scale, ranging from minimum to maximum possible expression).All observers were unaware of experimental treatments. Generalized Procrustes Analysis was used to generate consensus behavioural dimensions from these observations. Dimension one (46% of variance) was positively associated with terms such as 'exploratory' or 'confidant' and negatively with 'nervous' or 'unsure'. Dimension two (23% of variance) ranged from 'calm' to 'agitated'. Animal scores on these two dimensions were analysed using REML. In both tests Azaperone-treated pigs scored significantly higher (OF: W=19.66, P<0.001; EPM: W=34.98, P<0.001) on dimension one (i.e. they were observed as being more confidant/exploratory) than control pigs, with no treatment effect on dimension two (OF: W=0.00, P=0.978; EPM: W=0.16, P=0.696). Thus QBA detected the effects of an artificial manipulation of emotional state in pigs. This validation work supports the use of QBA as a research tool for the assessment of emotionality and welfare in animals.

**The welfare of pigs in five different production systems in France and Spain: assessment of behavior**

*Temple, D[1], Courboulay, V[2], Manteca, X[1], Velarde, A[3] and Dalmau, A[3], [1]UAB, UAB, 08193 Bellaterra, Spain, [2]IFIP, Institut du Porc, 35651 Le Rheu, France, [3]IRTA, IRTA, 17121 Monells, Spain; deborah.temple@uab.cat*

The 4th principle of the Welfare Quality® protocol labelled 'Appropriate Behavior' was assessed in pigs housed in 3 intensive systems (conventional in France and Spain, straw bedded in France, and Iberian intensive in Spain) as well as 2 Spanish extensive systems (Iberian extensive, Mallorcan Black pig). 60,263 pigs housed on 91 farms were evaluated over a 2 year period based on negative social and exploratory behavior, human animal relationship (HAR, panic was defined as > 60% pigs fleeing away from the observer) and qualitative behavior assessment (QBA). Multiple Generalized Linear Mixed Models were performed for negative social and exploratory behaviors as well as for the HAR test. First, models were built to compare the 5 production systems studied. Then, separate models were developed for each intensive system to identify possible predictive factors. Data from the QBA were analyzed by principal component analysis (PCA) and expressed at farm level. On the basis of the PCA, differences between systems were evaluated using General Linear Models. Pigs on conventional farms and Iberian intensive pigs showed the highest occurrence of negative social behavior (4.5%; 5.1%, respectively,P<0.05) while extensive Iberian pigs showed the lowest one (1.0%,P<0.05). Pigs on-straw presented a higher frequency of exploration (48.3%) and more panic (47.3%) than pigs in the conventional system (35.0%; 24.3%, respectively,P<0.05). However, no significant differences in exploration and panic response were observed among pigs in the conventional system, both intensive and extensive Iberian pigs and Mallorcan Black pigs. The scores of extensive farms on the 1st axis of the PCA from the QBA were higher (P<0.001) than those of intensive Spanish farms. A significant country effect (P<0.001) was observed when comparing the PCA scores of conventional farms. Finally, several predictive factors for negative social and exploratory behavior as well as for the panic response were identified in each intensive system. The results indicate that systems can be differentiated based on the occurrence of negative social behavior, and that the QBA is useful to distinguish systems within a country. However, explorative behavior is not sensitive enough to discriminate between intensive and extensive systems.

## The inter-observer reliability of qualitative behavioural assessments of sheep

*Phythian, Clare[1], Wemelsfelder, Francoise[2], Michalopoulou, Eleni[1] and Duncan, Jennifer[1], [1]University of Liverpool, Department of Epidemiology and Population Health, Leahurst, CH64 7TE, United Kingdom, [2]Scottish Agricultural College, Sustainable Livestock Systems, Edinburgh, EH26 0PH, United Kingdom; clare.phythian@googlemail.com*

Qualitative Behaviour Assessment (QBA) is a whole-animal methodology that assesses the expressive qualities of an animal's demeanour, or 'style' of behaving, using descriptors such as 'relaxed', 'anxious' or 'content'. QBA has been shown to be a reliable and feasible assessment technique in pigs, cattle, poultry, and other species. This study examined the inter-observer reliability of a fixed list of 12 QBA terms generated by sheep experts, and applied by experienced assessors from veterinary and farm assurance backgrounds. Assessors were 2 veterinary students and 4 veterinary surgeons (Group 1) and 7 farm assurance inspectors (Group 2). Groups met on different dates and were instructed to assess the same 12 video clips showing a range of sheep behavioural expressions, by scoring the 12 QBA descriptors (relaxed, dejected, thriving, agitated, responsive, dull demeanour, content, anxious, bright, tense, vigorous, distressed – presented for scoring in this order) on a Visual Analogue Scale. Assessor scores were analysed together and in separate groups using Principal Component Analysis (covariance matrix, no rotation). For the all-assessor analysis the first Principal Component (PC1, 49% of variation) ranged from 'content/relaxed/bright' to 'distressed/dejected/tense', while PC2 (31%) ranged from 'agitated/responsive/anxious' to 'dull/dejected/relaxed'. Analyses of separate groups were highly similar. Scores for individual assessors were correlated using Kendall's coefficient of concordance (W), giving values of 0.83 (PC1) and 0.84 (PC2) for all 13 assessors, 0.90 and 0.86 for Group 1, and 0.78 and 0.91 for Group 2 respectively. All values were significant at P<0.001. These results indicate that experienced assessor groups achieved excellent levels of inter-observer agreement using a pre-fixed QBA list to score sheep demeanour, and further support the reliability of QBA as a whole-animal assessment technique.

**Gregarious nesting as a response to risk of nest predation in laying hens**

*Riber, Anja B., Aarhus University, Dept. of Animal Health and Bioscience, Blichers Allé 20, Dk-8830 Tjele, Denmark; anja.riber@agrsci.dk*

Gregarious nesting occurs when a laying hen given the choice between an occupied and an unoccupied nest site chooses the occupied nest site. It is frequently observed in flocks of laying hens kept under commercial conditions, contrasting the behaviour displayed by feral hens that isolate themselves during nesting activities. What motivates laying hens to perform gregarious nesting is unknown. One possible explanation is that gregarious nesting is an antipredator response – not as an evolutionary adaptation in the traditional sense, but as a response to the risk of nest predation emerging from behavioural flexibility in nesting strategy. The present experiment aimed at investigating this hypothesis. Twelve groups of 14-15 Isa Warren hens age 44 weeks were housed in pens each containing three nest boxes. Nesting behaviour was recorded for 5 days in each of three distinct periods; a) pre-predator; a pre-exposure period, b) predator; a period with exposure to a daily simulated attack by a flying model of a hooded crow, and c) post-predator; a post-exposure period. Additional data collected were the behaviour of each hen 5 min prior to and 10 s after the simulated predator attacks. The proportion of gregarious nest box visits of the total number of visits, where the hens had a choice between gregarious or solitary nesting, was higher during the predator period ($P<0.01$). There was a tendency to an overall effect of period on number of visits to nest boxes ($P=0.08$); nest box visits were more frequent during the predator period than during the pre-predator period ($P<0.05$). The number of eggs laid in each nest box did not differ between periods ($P>0.05$), but lack of space during the sitting phase may have forced some hens to abandon occupied nest boxes and select unoccupied nest boxes for oviposition. The hens reacted with fear-related behaviour to the simulated predator attacks, e.g. fewer hens engaged in normal non-agitated behaviour after exposure to the predator model than before ($P<0.001$), and this did not change with day of exposure ($P>0.05$). In conclusion, some evidence was found for the proposed hypothesis that gregarious nesting is an antipredator response. However, knowledge about the cause of gregarious nesting is still sparse and until proved otherwise gregarious nesting should be considered as a behavioural activity influenced by multiple factors.

### Astroturf® as a dustbathing substrate for laying hens

*Alvino, Gina, Archer, Gregory and Mench, Joy, University of California, Davis, Department of Animal Science, One Shields Avenue, Davis, California, 95616, USA; gmalvino@ucdavis.edu*

Furnished cages for laying hens often contain an Astroturf® (AT) pad which may be sprinkled with feed to promote dustbathing. We evaluated hens given AT or AT plus feed to determine if these substrates are used for dustbathing. Laying hens (N=30) without prior exposure to substrate were housed singly in 91.4 cm × 45.7 cm × 45.7 cm cages at 34 weeks of age. Groups of 10 hens were randomly provided with either sand (SAND; control); an AT pad; or an AT pad with 200 g of feed (ATF) daily. Each hen was exposed to these substrates in a 3×3 Latin Square design, with each treatment period lasting 20 days. The treatment order was: SAND – ATF – AT; ATF – AT – SAND; AT – SAND – ATF. For each treatment period, behavior was recorded for 8 or 9 days, from 1100-2200 hours. Data were analyzed using the GLM or Kruskal-Wallis and Dwass-Steel-Critchlow-Fligner non-parametric tests. During treatment period 1 there were significantly fewer dustbathing bouts in SAND (mean = 3.11±0.48) than AT (7.30±1.35; P=0.034) and fewer bouts on the wire floor (1.29±0.61) than both ATF and AT (3.97±0.81 and 6.52±1.36, respectively; P=0.004). Hens in SAND also spent less time dustbathing on wire (mean = 1.94±0.80 min) than ATF and AT (10.44±2.21 and 14.88±3.65, respectively; P=0.002) and more time dustbathing in the substrate (mean = 16.83±2.63) than ATF and AT (4.91±2.26 and 2.97±1.34, respectively; P=0.0002). During treatment period 2, however, there were no differences in bout number, but SAND spent more time dustbathing in substrate (mean = 12.59±4.15) than AT (1.68±1.29; P=0.0002). During treatment period 3, preliminary analysis showed that hens in SAND performed fewer bouts overall (mean = 4.24±0.78) as well as fewer bouts on the wire floor (3.75±0.85) than ATF (7.88±0.77 and 7.84±0.75, respectively; P=0.03, 0.02, respectively), but there were no treatment differences in total time spent dustbathing. The findings suggest that AT does not provide an adequate dustbathing substrate even with feed and that exposure to AT or ATF as a dustbathing substrate may even be aversive to hens, since across treatment periods the proportion of bouts in sand decreased from 0.59 to 0.25 to 0.11 and the proportion of bouts on wire increased.

**Feather pecking and serotonin: 'the chicken or the egg?'**

*Kops, Marjolein S.[1], Bolhuis, Elizabeth J.[2], De Haas, Elske N.[2], Korte-Bouws, Gerdien A.H.[1], Rodenburg, T. Bas[2], Olivier, Berend[1] and Korte, S. Mechiel[1], [1]Utrecht Institute for Pharmaceutical Sciences, Universiteitsweg 99, 3584 CG Utrecht, Netherlands, [2]Adaptation Physiology Group, and Animal Breeding and Genomics Centre, P.O. Box 338, 6700 AH Wageningen, Netherlands; m.s.kops@uu.nl*

Feather pecking (FP) is a detrimental behavior causing welfare problems in laying hens. Given that only a few individuals initiate FP in a flock, it is worth investigating neurobiological characteristics of hens that develop FP. Aim of this study was to analyze brain monoamines in laying hens from different lines and displaying phenotypic differences. Hens of 33 weeks were sacrificed after a 5-min manual restraint test. Effects of genetic line (control or low mortality line, fourth generation) and of behavioral phenotype (pecker, non-pecker or victim) on monoamine levels in five brain areas were determined using General Linear Models. Dopamine ((DOPAC+HVA+3-MT)/DA) and serotonin turnover levels (5-HIAA/5-HT) did not differ between lines (low mortality, n=20 vs. control, n=20) or between pecker (n=9), non-pecker (n=9) and victim (n=9) in the medial striatum, amygdala, caudolateral part of nidopallium, and hippocampus. Peckers had higher serotonin turnover in the thalamus than non-peckers, with levels of victims in between (phenotype effect, $P<0.05$). Moreover, hens from the low mortality line showed lower NA, DOPAC and HVA levels ($P<0.05$) and tended to show lower DA and 5-HIAA levels ($P<0.10$) in the amygdala than control line hens. To conclude, line differences in monoamine levels in the amygdala fit the idea that selection against mortality influences fearfulness in chickens. Remarkably, FP in adult hens was not associated with low 5-HT turnover levels; this in contrast to previous studies. It is likely that low brain serotonin levels are involved in active coping, while high levels could lead to passive coping. Our new hypothesis is that active 'Hawk-like' animals initiate feather pecking (first order FP), whereas passive 'Dove-like' animals (being more aware of changes in the environment, e.g., feather damage) learn to feather peck once there is feather damage (second order FP). Therefore we think it is crucial to differentiate between first and second order feather peckers when studying the relationship between brain monoamine levels and FP.

## Effects of predictability on feeding and aversive events in captive rhesus macaques (*Macaca mulatta*)

*Gottlieb, Daniel H.*[1,2], *Coleman, Kristine*[3] *and Mccowan, Brenda*[1,4], [1]*California National Primate Research Center, One Shields Avenue, Davis, CA 95616, USA,* [2]*University of California, Davis, Animal Behavior Graduate Group, One Shields Avenue, Davis, CA 95616, USA,* [3]*Oregon National Primate Research Center, 505 NW 185th Ave, Beaverton, OR 97006, USA,* [4]*University of California, Davis, School of Veterinary Medicine, Department of Population Health and Reproduction, One Shields Avenue, Davis, CA 95616, USA; dhgottlieb@ucdavis.edu*

Rhesus macaques housed indoors experience many husbandry activities on a daily basis. The anticipation of these events can lead to stress, regardless of whether the events themselves are positive or aversive. Previous research suggests that making daily events highly predictable will decrease stress and improve welfare. However, some studies have found conflicting results regarding the effects of predictability on welfare. Thus, it is imperative that we empirically test the effect of increasing predictability before implementing new practices for a given species. The specific goal of this study was to identify whether increasing the predictability of daily feeding and cleaning events could decrease stress and anxiety in captive rhesus macaques. This study was conducted on 39 singly housed subjects in four rooms at the Oregon National Primate Research Center (ONPRC). Current daily routines at the ONPRC were modified to include temporal predictability, signaled predictability, or both. Temporally predictable events occurred reliably at the same time daily, while signaled predictable events were preceded by a distinct event-specific signal in the form of a doorbell. Each subject received all four conditions: unpredictable events, temporally predictable events, signaled predictable events, and temporally and signaled predictable events. The order of events was balanced using a Latin square design. Stress and anxiety under each condition were evaluated by expression of stereotypic and displacement behaviors. Data were analyzed using generalized mixed effects modeling with individuals as the experimental unit and room accounted for as a random effect. Our results showed that feeding events elicited less stress and anxiety behaviors when temporally predictable. In contrast, stress behaviors did not always decrease when events were preceded by the event-specific signal, which increased stress behaviors for some events.

**Exercise pens as an environmental enrichment for laboratory rabbits**

*Lidfors, Lena[1], Knutsson, Maria[1], Jalksten, Elisabeth[2], Andersson, Håkan[2] and Königsson, Kristian[2], [1]Swedish University of Agricultural Sciences, Department of Animal Environment and Health, P.O. Box 234, SE-532 23 Skara, Sweden, [2]AstraZeneca R&D, Safety Assessment Sweden, Gärtuna, SE-151 85 Södertälje, Sweden; Lena.Lidfors@slu.se*

Laboratory rabbits are usually kept singly in cages where running and environmental enrichment is restricted. The aim was to study differences in behavior and corticosterone in male New Zealand White rabbits with or without access to exercise pens. 21 rabbits (4 months–3 years old) were singly housed in cages with stainless steel walls and perforated polypropylene floors ($1.04 m^2$, 0.65 m high) with shelves, wooden chew sticks and 24 h access to hay and water. The animals were divided into 3 groups: control or allowed to exercise either 1 or 3 times weekly. Animals in exercise groups had access to pens individually for one h in day time during 8 weeks. Three exercise pens ($3.6 \times 0.9 m^2$, 0.8 m high) made in plastic coated steel wire mesh were placed in the middle of the animal room. Each pen contained rubber flooring, a shelf, a box with wood shavings, hay, a water bottle, a wooden chew block and a plastic ball. Behaviors of all groups were recorded instantaneously at one min intervals during 60 min both in cages and pens. Blood samples for corticosterone were taken from an ear vessel before and at 1st, 4th and 8th week after exercise. Rabbits were weighed weekly. A linear statistical model was fitted to the behavioral data and a pair wise t-test was used for the corticosterone. Moving was the most common behavior in the pens (50% of obs.), and higher than in the cages (5%, P<0.05). Lying was most common in the cages (50%), whereas in the pens it was shown 13% (3 times exercise) vs. 5% (1 time exercise, n.s.). Sitting was more common in pens (16%) than in cages (9-13%, P<0.05), whereas grooming was more common in cages (9-11%) than in pens (5-8%, P<0.05). Eating did not differ between pens (5-8%) and cages (13-20%, n.s.). Hiding occurred at only 0.5-1% of recordings, and mainly during the first exercise session. Corticosterone was elevated after exercise the first week compared to the week before exercise (P<0.05), but not during week 4 and 8. Rabbits exercised lost some weight during the exercise period compared to control rabbits. In conclusion access to exercise pens activated the rabbits and caused a transient elevation of the corticosterone levels.

## Does the presence of a human affect the preference of enrichment items in young isolated pigs?

*Deboer, Shelly[1], Garner, Joseph[1], Eicher, Susan[2], Lay Jr., Donald[2], Lucas, Jeffery[1] and Marchant-Forde, Jeremy[2], [1]Purdue University, 125 S Russell St, W Lafayette, IN 47907, USA, [2]LBRU USDA-ARS, 125 S Russell, W Lafayette, IN 47907, USA; slpfeffe@purdue.edu*

Pigs may be housed individually in both production and research settings. Gregarious by nature, pigs kept in isolation may show behavioral and physiological signs of stress. The aim of our study was to determine the preference of individually-housed pigs, for social and non-social enrichments. Three enrichment items were compared: a mat on part of a woven wire floor (MAT), a companion visible across the passageway (COM) and a mirror on one wall (MIR). Weaner pigs (Yorkshire × Landrace, N=14) were housed individually with continuous access to 4 adjacent pens (1.5 x 3.0 m), 3 of them containing one enrichment and one control (C) pen with no enrichment. All pens had equal access to feed and water. The animal was only able to access each enrichment item while in that enrichment's pen. Pigs were video recorded 14 h/day for 7 days. Videos were analyzed by scan sampling every 10 minutes to determine location, posture and behavior. Differences in the enrichment preference of the pigs were tested using a GLM model in JMP. Data are presented as mean percentage of time with 95% confidence interval. Pigs spent more time (P<0.05) in the COM pen (36.2% (23.7-49.7%)) compared to C (9.7% (3.2-19.2%)) with MAT (29.3% (17.8-42.4%)) and MIR (16.4% (7.7-27.7%)) as intermediates. A second analysis was performed on the data to investigate changes in preferences in the presence or absence of a human in the room. The pens were then combined into 2 categories: social pens (COM and MIR) and nonsocial pens (MAT and C). These data were analyzed using Proc Glimmix in SAS. The probability of a pig choosing social pen when a human was present (0.8967) was significantly higher (P<0.001), then when absent (0.5243). Within the social enrichments, the probability of the animal choosing either MIR or COM was not different. Our results confirm that preference studies are highly sensitive to experimental conditions and the assumption that the most important preference is the one the animal spends most of its time with can be misleading. It appears that a mirror may be used by the animal for social support during periods of perceived stress, however further investigation is warranted.

**Playful handling before an intra-peritoneal injection induces a positive affective state in laboratory rats**

*Cloutier, Sylvie, Wahl, Kim, Panksepp, Jaak and Newberry, Ruth C., Washington State University, Department of VCAPP, CSAW, P.O. Box 646520, Pullman WA, USA, 99164-6520, USA; scloutie@vetmed.wsu.edu*

We hypothesized that the timing of playful handling (tickling) in relation to an intra-peritoneal injection affects the efficacy of tickling in reducing stress associated with the procedure. Male Sprague-Dawley rats (N=96) were either injected with saline intra-peritoneally (I) or handled similarly but not injected (control, C), and exposed to one of two handling treatments: not handled (N) or tickled (T) for 2 min immediately before injection (B), after injection (A) or both before and after injection (BA), resulting in 8 treatment groups (IN, ITB, ITA, ITBA, CN, CTB, CTA, CTBA). Treatments were administered daily for 10 days from 32-41 days of age. We compared the rate of frequency-modulated 50-kHz ultrasonic vocalizations (USVs) (associated with positive affective states) emitted before and after injection, and duration of injection on treatment days 1 and 10. Overall, tickled rats emitted more USVs than N rats. Before injection, the call rate of rats in the 'tickled after' groups (ITA, CTA) did not differ from that of IN and CN rats. After injection, rats in the 'tickled before' groups (ITB, CTB) emitted more calls than IN and CN rats, and ITBA rats emitted more calls than ITA rats (Median (IQR) calls/min, ITB: 52 (26-69), CTB: 74 (25-108), ITBA: 172 (123-197), ITA: 119 (37-170), IN: 0 (0-6), CN: 1 (0-9); Mixed Model Anova on ranked data, F(7.88)=40, P<0.0001). Emission of USVs was higher on day 10 than day 1 both before and after injection (Before: day 1: 14 (0-82) calls/min, day 10: 84 (19-158); F(1.88)=185, P<0.0001; After: day 1: 44 (2-116), day 10: 112 (45-169); F(1.88)=115, P<0.0001). Rats tickled before, and before and after, injection required less time to inject than non-handled rats (ITB: 10 (9-12) s, ITBA: 10 (9-12.5); IN: 13.5 (10.5-20); F(7.88)=17, P<0.0001). Our results suggest that tickling before, or before and after injection, is effective in inducing a positive affective state in rats and mitigating the aversiveness of injections.

## The naked truth: breeding performance in outbred and inbred strains of nude mice with and without nesting material

*Gaskill, Brianna N[1], Winnicker, Christina[2], Garner, Joseph P[1] and Pritchett-Corning, Kathleen R[2], [1]Purdue Univ., 915 W. State St, West Lafayette IN 47906, USA, [2]Charles River, 251 Ballardvale St, Wilmington MA 01887, USA; christina.winnicker@crl.com*

In laboratories, mice are housed at ambient temperatures between 20-24 °C, which is below their lower critical temperature of 30 °C, but comfortable for human workers. Thus, mice are thermally stressed, which can compromise many aspects of physiology from metabolism to pup growth. These effects may be exacerbated in the breeding of nude mice. We hypothesized that nesting material would allow nude mice to behaviorally thermoregulate, reducing heat loss to the environment. We predicted this reduction will improve feed conversion as well as breeding performance. We housed naïve Crl:NU-Foxn1[nu] and CAnN.Cg-Foxn1[nu]/Crl breeding trios (2 haired females:1 nude male; 30 cages per strain) in shoebox cages at ≈21 °C either with or without 8 g nesting material for 6 months within an isolator. Nest quality was scored weekly using a previously published standard scale. Feed was weighed when added and weighed back at the end of the experiment. At weekly cage changes fresh nesting treatment was provided. Reproductive observations were made three times a week and pups were weighed and sexed at weaning (21-28 days). Analyses used GLMs with post-hoc contrasts. Nesting material significantly increased the number of pups weaned per cage ($F_{1,55}=12.44$; $P<0.001$; 29.6±2.1 vs 19.6±1.9). The amount of feed needed to produce 1 g of weaned pup was nearly halved when mice were provided nesting material ($F_{1,55}=8.5$; $P=0.005$; 17.5 g±3.8 vs 33.5 g±3.9). However the total feed consumed by both treatments was not significantly different ($F_{1,53}=1.58$; $P=0.21$). The breeding index (pups weaned/female/week) was significantly higher when nesting material was provided ($F_{1,55}=10.15$; $P=0.002$; 0.64±0.04 vs 0.44±0.04). Thus nests lessen the thermal impact of standardized cool temperatures on nude mice. However, the energy (using feed consumption as a proxy) conserved by nesting material is not simply freed up from heat generation but reallocated to improved breeding performance. Together these data show that good welfare is good business and good science.

**Beneficial effects of environmental enrichment on emotional reactivity of Japanese quail submitted to repeated negative stimulations**

*Laurence, Agathe[1], Houdelier, Cécilia[1], Petton, Christophe[1], Calandreau, Ludovic[2], Arnould, Cécile[2], Favreau-Peigné, Angélique[2], Boissy, Alain[3], Leterrier, Christine[2], Richard-Yris, Marie-Annick[1] and Lumineau, Sophie[1], [1]Université de Rennes, UMR 6552, 263 avenue du Général Leclerc, 35042 Rennes cedex, France, [2]INRA, UMR 85 Physiologie de la Reproduction et des Comportements, Nouzilly, 37380 Nouzilly, France, [3]INRA, Unité de Recherche sur les Herbivores, Theix, 63 122 St Genès Champanelle, France; agathe.laurence@univ-rennes1.fr*

We investigated interactions between environmental enrichment (EE) and repeated negative stimulations (RNS) during the ontogeny of Japanese quail. Half of the treated individuals (set T, N=30) and half of the control animals (set C, N=28) were housed in enriched cages (set E), whereas the other halves were housed in traditional cages (set NE). Treated individuals were submitted to RNS for two weeks (4 to 5 stimulations per 24 h). All subjects were observed in their home cages during the stimulation procedure, and then were presented emotional reactivity tests. When data were normally distributed, analysis was performed using a two-way ANOVA which examined the main effects of EE and RNS and their interactions; otherwise data were analyzed with Mann-Whitney U-tests for each factor independently. During the stimulations procedure, T quail preened less frequently than C quail, independently of cage type (10th day: set C: 13.10±0.02%, set T: 8.8±0.01%, ANOVA, RNS effect, P=0.01). E-T quail were observed more frequently in the rear zone of their cage than were E-C quails when the experimenter was in sight (10th day: set C: 11.4±3.3%, set T: 31.4±7.6%, Mann-Whitney U-test: P=0.03). The stimulation procedure affected the emotional reactivity of the quail in relation to cage type: E-T quail were less reactive than were NE-T quail (i.e. number of steps in open-field test, set NE-T: 66±21, set E-T: 138±32, Mann-Whitney U-test: P=0.05). We evidenced an interaction between the two tested factors: negative stimulations inhibited dust bathing in NE quails whereas they enhanced it in E quails (i.e. dust bathing latency in emergence test, ANOVA, interaction between EE and RNS: P=0.003, set NE-T: 139±12 s, set E-T: 78.5±11 s, set NE-C: 92±15 s, E-C: 113±15 s). Thus, environmental enrichment had a positive effect on emotional reactivity when quails were exposed to repeated negative stimulations, but it had little effect on control quails.

**Effects of conditioning on blood draw in cats**

*Lockhart, Jessica, Wilson, Karri and Lanman, Cindy, P&G Pet Care, 6571 SR503N, Lewisburg, OH 45338, USA; lockhart.j.4@pg.com*

We measured the impact of operant training to accept jugular blood draws in a recumbent (novel) position by domestic cats. Cats were assigned to one of three groups: G1 (N=13): no training, traditional jugular blood draw; G2 (N=17): trained in novel position but traditional blood draw used; G3 (N=14): trained in novel position and novel blood draw used. The impact of handler was gauged by testing each cat twice, once with a familiar and once with an unfamiliar person, one week apart; random sampling order each day. For each test, cats received two blood draws 20 minutes apart. Blood samples were analyzed for cortisol levels with draw 1 serving as the initial stressing event/baseline and draw 2 serving as test/change from baseline. All blood draws were filmed and coded for behavioral signs of stress. Cats displayed significantly more escape attempts with the unfamiliar ($1.23\pm0.20$) than the familiar handler ($0.49\pm0.20$), Mixed linear model ANOVA, $F(1, 99.87) = 14.65$, $P<0.00$. There was a significant effect for group and time to position ($F(2, 30.97)=6.76$, $P<0.00$). Paired comparison with Bonferroni adjustments showed that G3 ($46.53$ s$\pm4.66$) took significantly longer to position than G2 ($24.75\pm3.67$, $P<0.00$) and G1 ($33.26\pm4.12$, $P<0.04$), but overall took the same amount of time to complete blood draws. There was a significant difference between heart rates at release between groups, $F(2, 17.66) = 8.99$, $P<0.00$. G3 had lower heart rates when released ($178.79\pm7.22$) than G2 ($217.38\pm4.91$, $P<0.00$) and G1 ($208.39\pm4.99$, $P<0.00$). This suggests that the G3 cats showed the least physiological reaction to the blood draw. Trained cats, despite method or familiarity with handler, showed lower cortisol levels when the procedure was repeated. Cortisol levels did not differ significantly at baseline on either day ($P>0.05$) or between groups on day 1 ($P>0.05$). However, there was a significant difference in cortisol levels between groups on day 2, $F(2, 21)=4.703$, $P<0.021$ where G1 ($2.656\pm0.633$) had significantly higher test cortisol concentrations when compared with G2 ($0.354\pm0.540$, $P<0.01$) or G3 ($0.134\pm0.801$, $P<0.02$). In conclusion, operant training to blood draws appears to have a positive impact on the cat's experience whether a traditional or recumbent position is used. These results support the use of operant training to improve the overall blood draw experience for domestic cats.

## Is hair and feather pulling a disease of oxidative stress?

*Vieira, Giovana, Lossie, Amy, Ajuwon, Kola and Garner, Joseph, Purdue University, Animal Sciences, 915 West State St., 47906, West Lafayette, IN, USA; gvieira@purdue.edu*

Barbering is an abnormal repetitive behavior commonly seen in laboratory mice, where a 'barber' mouse plucks hair from its cagemates or itself, in idiosyncratic patterns, leaving the cagemates with patches of missing fur and/or whiskers. Several lines of evidence validate barbering as a model of Trichotillomania (human compulsive hair pulling), and barbering may also model hair and feather pulling welfare issues in other species. N-Acetylcysteine, (NAC) a cysteine derived food additive, is remarkably effective in treating Trichotillomania patients, but its mechanism of action is unknown. Reactive Oxygen Species (ROS), also known as free radicals, form as a natural byproduct of the normal metabolism of oxygen, which in turn is regulated by the Hypothalamic-Pituitary-Adrenal (HPA) and Sympathetic-Adrenal-Medullary (SAM) axes. Thus a variety of factors, including diet and chronic stress, elevate ROS in the body. Under normal circumstances, cells are able to defend themselves against ROS damage with antioxidant pathways. Nerve cells are particularly vulnerable to oxidative damage from ROS, and NAC is the precursor to the main antioxidant produced to defend the brain. We therefore hypothesized that barbering is caused by neuronal damage or quiescence as a result of multifactorial sources for elevated ROS, and/or a failure to activate antioxidant defenses. We tested this hypothesis in 26 female C57BL/6J mice aged between 2 and 8 months of age selected from our colony. We identified mice as barbers or non barbers, and collected a minimum of 0.5 ml of urine from each mouse. Urine was analyzed for total antioxidant capacity and creatinine (to control for urine concentration). We used logistic regression to test whether total antioxidant capacity:creatinine ratio was a predictive biomarker for barbering. The analyses were blocked by cage. We found that barbers had higher total antioxidant capacity of urine than non-barbers (LR Chi Sq=10.4; P=0.0013). This is consistent with a failure to activate antioxidant defenses, confirming a relationship between oxidative stress and barbering behavior, and providing a potential physiological biomarker for the disease mechanism.

## Implications for animal welfare: habituation profiles of 129S2, 129P2 and 129X1 mouse strains

*Boleij, Hetty[1,2], Salomons, Amber R.[1,2], Arndt, Saskia S.[1,2] and Ohl, Frauke[1,2], [1]Faculty of Veterinary Medicine, Animals in Science and Society, Yalelaan 2, 3584CM Utrecht, Netherlands, [2]RMI, Universiteitsweg 100, 3584CGUtrecht, Netherlands; h.boleij@uu.nl*

An animal should adapt to a novel environment, showing less anxiety over time (habituation). Welfare might be compromised when anxiety becomes non-adaptive. Genetic factors can play a role in non-adaptive anxiety. Based on previous work in which 129P3/J mice showed increased anxiety-related behavior over time, four other 129 substrains were tested on habituation in the modified hole board (mHB) test. The mHB consists of a protected (near side walls) and an unprotected area (hole board in center of arena). Avoidance of the unprotected area is indicative of anxiety. Male mice of the 129S2Pas, 129S2Hsd (experiment 1) and 129P2 and 129X1 (experiment 2) strains were tested (n=8 per strain) 4 trials (5 min per trial) per day, for 5 days. Among others, anxiety-related, exploratory and locomotor behavior was observed. Within strain effects over trials were used as indication of habituation (decrease over time) or sensitization (increase over time). Strain comparisons were done per experiment. A RM-ANOVA was performed within experiments. All 129 strains showed increased avoidance of the unprotected area over trials (trial effects $P<0.01$) indicated by an increased latency (s) to enter (129X1: from 75.0±19.1 to 251.4±24.1; 129P2: from 34.7±12.3 to 205.1±25; 129S2Pas: from 199.7±36.0 to 235.4±17.5;129S2Hsd: from 132.2±27.8 to 276.6±13.6) and a decreased time (s) spent in the unprotected area (129X1: from 31.5±11.3 to 0.7±0.5; 129P2: from 42.1±9.7 to 2.7±1.3; 129S2Pas from 25.6±1.9 to 1.9±1.1; 129S2Hsd 47.9±9.6 to 0.8±0.4), the mentioned values are means±SEM from trial 1 and trial 20. Strain differences were observed in locomotor (strain, strain*trial and trial effect $P<0.01$) and exploratory activity (experiment 1: latency and number of rearings: strain*trial effect $P<0.01$, experiment 2: number of hole explorations: strain*trial effect $P<0.01$). In conclusion, mice from the 129 substrains show increased anxiety over time, suggesting an inability to adapt to novel situations. This finding is of high importance considering housing and treatment conditions. It might be necessary to adapt those (e.g. lower frequency of cage changes, longer acclimatization periods) for different (sub-) strains of mice in order to benefit their welfare.

## Identification methods in newborn C57BL/6 mice: a developmental and behavioural evaluation

*Castelhano-Carlos, Magda João[1], Sousa, Nuno[1], Ohl, Frauke[2] and Baumans, Vera[2], [1]Life and Health Sciences Research Institute (ICVS), University of Minho, Campus de Gualtar, 4710-057, Portugal, [2]Faculty of Veterinary Medicine, Utrecht University, Department of Animals, Science and Society, P.O. Box 80166, 3508 TD Utrecht, Netherlands; mjoao@ecsaude.uminho.pt*

Few studies have been designed to assess possible negative effects of individual identification methods of newborn mice. In the present study three different identification methods were applied to newborn C57BL/6J mice on postnatal day (pnd) 5: toe clipping (Tc – 11 females and 11 males), toe tattoo ink puncture (Tip – 10 females and 10 males) and subcutaneous implantation of a small transponder (ScT – 8 females and 8 males); 10 males and 10 females were used as controls. The transponder used was indicated for newborn mice from the age of pnd3-5 onwards and was 0.24% of the weight and ¼ the length of the newborns used in this study. No mortality was observed in consequence of any of the identification methods applied.Newborn mice showed the least reaction to Tc followed by TIp and ScT, as observed by theirsudden movements and presence or absence of vocalization and urination as indicators of pain/ distress (Fisher's exact test, P<0.05). Importantly, clipped toe tissue proved to be enough for genotyping. No overall consistent differences in somatic and neurological reflex development during the postnatal period were shown as a result of the newborn individual identification. Further, none of the methods interfered significantly with the adult animals' general normal behavior (including e.g. grasp and climb) and sensory-motor functions as assessed with a simplified SHIRPA battery of tests, as well as Rotarod and EPM tests (P≤0.05). Post-mortem thymus and adrenal glands weights gave no indication of chronic stress as a consequence of the identification method. We conclude that toe clipping might even be advisable in newborn mice at a very young age, when genotyping is needed, contributing for early colony management and for improving welfare while avoiding other biopsies. Toe tattoo ink puncture is also a good identification method as it was shown to cause minimal discomfort when applied to the newborn and it is easy to observe in adult animals without need for handling. Transponder implantation should only be used in older newborns.

## Behavioral and physiological thermoregulation in mice with nesting material

*Gaskill, Brianna N[1], Gordon, Christopher J[2], Davis, Jerry K[1], Pajor, Edmond A[3], Lucas, Jeffrey R[1] and Garner, Joseph P[1], [1]Purdue Univ., 915 W. State St., West Lafayette, IN 47907, USA, [2]Environmental Protection Agency, Research Triangle Park, NC 27709, USA, [3]Univ. of Calgary, 3280 Hospital Dr. NW, Calgary T2N 4N1, Canada; bgaskill@purdue.edu*

In laboratories, mice are housed at ambient temperatures ($T_a$) between 20-24 °C, which is below their lower critical $T_a$ of 30 °C. Thus, mice are thermally stressed, which can compromise many aspects of physiology from metabolism to behavior. These changes reflect impaired wellbeing and can possibly affect scientific outcomes. We hypothesized that nesting material would allow mice to alleviate cold stress by controlling their thermal microenvironment within the cage. We predicted that nesting material will insulate the mice, reduce heat loss, and decrease non-shivering thermogenesis. We housed naïve C57BL/6, CD-1, and BALB/c mice (24 males and 24 females/strain in groups of 3) in standard cages at 20 °C either with or without 8 g nesting material for 4 weeks. Nests were scored daily. Thermal properties of the nests were assessed once a week using a thermal imaging camera. Body core temperature ($T_b$) was followed using radio telemetry from one mouse per cage. During weekly cage cleanings fresh nesting treatment was provided. Scapular brown fat was analyzed for expression of UCP-1, a protein produced in thermogeneration, by qPCR. Analyses used GLMs with post-hoc contrasts. Nesting material was more insulating ($P<0.05$) and the mean radiated temperature was negatively correlated with nest score ($P<0.05$). Thus, higher nest scores resulted in less radiated heat. No treatment effects on $T_b$ were found ($P>0.05$). CD-1s with nesting material had higher end body weights than controls ($P<0.05$). No effect was seen in the other two strains. Mice with the telemetry implant had larger spleens than controls, indicating an immune response to the implant. Balb/c mice with nesting material express less mRNA for the UCP-1 protein than controls ($P<0.05$). This indicates that Balb/c's with nesting material do not utilize their brown fat to create heat as readily as controls. Nests can alleviate thermal discomfort by decreasing the amount of radiated heat and reduce the need for non-shivering thermogenesis. However, different strains appear to use different strategies to maintain a constant $T_b$ under cool standard laboratory $T_a$.

**Tail biting alters feeding behavior of victim pigs**

*Viitasaari, Elina[1], Hänninen, Laura[1], Raekallio, Marja[2], Heinonen, Mari[1] and Valros, Anna[1], [1]Research Centre for animal welfare, University of Helsinki, Finland, Department of production animal medicine, Koetilantie 7, PL 57, 00014 University of Helsinki, Finland, Finland, [2]Faculty of veterinary medicine, University of Helsinki, Finland, Department of equine and small animal medicine, Koetilantie 7, PL 57, 00014 University of Helsinki, Finland, Finland; elina.viitasaari@helsinki.fi*

Little is known about the effect of tail-biting on the feeding behavior of victim pigs. Our hypothesis was tail-biting to decrease feeding related parameters. Therefore, in this longitudinal cohort study we studied automatically collected feeder data from a finishing herd with one automatic one-space feeder per group of 11 pigs. We selected 13 tail-bitten pigs with fresh bite wounds weighing 30-90 kilograms in 7 pens and observed the data from 5 days before to 5 days after the bite wound was first noticed at day 0. We calculated mean daily duration at feeder, mean daily intervals between feeder visits and mean daily feeding efficiency (amount of feed consumed in grams divided by time spent at feeder in seconds). The differences between the observed days were compared with repeated measures mixed models thus the pigs served as their own controls. The mean ±se time spent in feeder, feeding efficiency and feeder visit intervals differed significantly between days ($P<0.001$ for all). The duration in feeder decreased from 64.18±9.23 min to 56.62±9.23 min between days -1 to day 0 and increased again to 63.66±9.23 min until day 2 ($P<0.05$). Feeding interval was 27.74±12.17 min on day -1 and increased from 22.87±12.17 min on day 0, to 43.80±12.17 min on day 2 ($P<0.05$). Feeding efficiency elevated from 0.62±0.07 g/s to 0.66±0.07 g/s from day -1 to day 2 ($P<0.05$). No significant differences were present in any parameters before tail biting, from days -5 to -2. Tail-biting altered feeding behavior of victims in this study. Observed change in feeding efficiency, increase in feeding interval and decrease in time spent in the feeder suggest restlessness of the victim pigs during the outbreak, which can be related to pain. According to our findings, alterations in feeding behavior before bite wound was first noticed might be more sensitive indicator of tail biting compared to visual inspections by human observer.

**The effects of diet ingredients on gastric ulceration and stereotypies in gestating sows**

*Wisdom, Stephanie L.[1], Richert, Brian T.[1], Radcliffe, J. Scott[1], Lay Jr., Donald C.[2] and Marchant-Forde, Jeremy N.[2], [1]Purdue University, 915 W. State St., W. Lafayette IN 47907, USA, [2]USDA-ARS-LBRU, 125 S. Russell St, W. Lafayette IN 47907, USA; slrtfc@me.com*

Stereotypies in swine can be altered with feedstuffs, but it is unknown how this will affect the development of gastric ulcers. The objective of this experiment was to determine the effects of omeprazole and sodium bicarbonate on ulcerations of the pars esophagea (UPE) of the stomach and oral behavior in gestating sows. Thirty-six stall-housed sows were assigned to 1 of 3 treatments with parities (1.68±0.22) balanced across treatments. Treatments were: (1) control, corn-SBM based diet; (2) omeprazole, diet 1 + a single dose of 60 mg omeprazole; and (3) sodium bicarbonate, diet 1 + 2% sodium bicarbonate. Treatments began d30 of gestation and diets were fed once per day at 2.05 kg/d. Sows underwent endoscopic evaluation at d30 to assess initial UPE. UPE was also investigated at d60 and d90 of gestation. Evaluation was by a trained veterinarian using an endoscope while sows were anesthetized. Ulcers were scored using a 7-point scale, ranging from 0, no visible lesions, to 6, deep ulcerations in >20% of the pars esophagea. Sow behavior was recorded for 30 min/d starting 30 min after feeding, 1 and 4 wk after starting treatments to determine frequencies and durations of stereotypies. Heart rates (HR) were recorded simultaneously to determine HR response and thus feeding motivation. All data were analyzed using the Mixed procedure of SAS with treatment, parity, day, HR, UPE and behavior as factors. UPE differed between groups before treatment was applied (P<0.01), but UPE was not balanced initially because no difference was anticipated. Using initial UPE (d30) as a covariate, there were no effects of treatment on d60 or d90 UPE (both P>0.05). The average UPE score was 1.21±0.28 ranging from 0 to 6. HR increased from 76.2±1.9 beats per min (BPM) before feeding to 110.6±3.0 BPM after feeding (P<0.001). There was no effect of treatment, parity, or test day on HR (P>0.05). While treatment had no effect on behavior (P>0.05), there was an effect of d60 UPE on bar-biting, eating, and drinking (P<0.01) with eating decreasing and bar-biting and drinking increasing as ulcers worsen. Further studies are needed to evaluate the effects and UPE treatments on sow behavior.

## Energy balance and feeding motivation of sheep in a demand test

*Doughty, Amanda[1,2], Hinch, Geoff[2], Ferguson, Drewe [1] and Matthews, Lindsay [3], [1]CSIRO, Armidale, NSW, 2350, Australia, [2]University of New England, Armidale, NSW, 2351, Australia, [3]AgResearch, Hamilton, 3240, New Zealand; Amanda.Doughty@csiro.au*

The measurement of strength of motivation has become an important tool used to assess the resources that an animal values and, subsequently, to aid in determining its welfare. However, the relationship between animal motivation and welfare state is not well defined with conclusions based on the assumption that welfare is reduced if a 'valued' resource is not provided. One way to better identify the relationship between the level of motivation and welfare state is by investigating the mechanisms that affect motivation in a demand test. This study therefore aimed to test the hypothesis that energy balance (energy expended = energy consumed) could determine the motivation to work for food and ascertain if this balance could be altered by gut storage levels. A secondary aim was to test the hypothesis that two measures of demand ($P_{max}$ or point of maximum work, and maximum price paid (MPP) or point of work ceasing) would be equally responsive to changes in energy balance. Sheep were tested individually (with companions nearby) to see how many times in a 23 h period they would walk a specific distance for a 5 g (55 kJ) food reward. Eight sheep were trained in a 50 m U-shaped laneway to access a double-sided feeder and gained a reward with each access event. The distance (cost) that the sheep walked was increased progressively on a log scale (1.5-105.5 m). Sheep were allocated to one of two treatments (14-h restriction and an unrestricted control) in a cross-over design. Data was analysed using ANOVAs and the demand function was calculated as $Ln (Q) = In(L) + b[In(P)] - a(P)$. The results showed $P_{max}$ was similar for both treatments (P>0.05), however, MPP could not be calculated as sheep did not stop walking at the maximum distance tested. The cost at zero energy balance was also similar for both treatments (P>0.05) and the sheep continued walking beyond this point. The cost at zero energy balance was not significantly different from $P_{max}$ (P>0.05). These results indicate that in a demand test energy balance may, to some extent, determine the motivation to work for food and also that the amount of energy expended / energy consumed was not significantly altered by a short-term food restriction.

## Effect of milk feeding level on the development of feeding behaviour patterns in dairy calves

*Miller-Cushon, Emily K.[1], Bergeron, Renée[2], Leslie, Ken E.[3] and Devries, Trevor J.[1], [1]University of Guelph, Kemptville Campus, Animal and Poultry Science, 830 Prescott Street, Kemptville, ON, K0G 1J0, Canada, [2]University of Guelph, Alfred Campus, Animal and Poultry Science, 31 St. Paul Street, Alfred, ON, K0B 1A0, Canada, [3]University of Guelph, Population Medicine, 50 Stone Road East, Guelph, ON, K0G 1J0, Canada; emillerc@uoguelph.ca*

There is evidence that early experiences may influence development of characteristic feeding patterns in dairy cattle. The objective of this study was to determine how milk feeding level affects development of feeding patterns in dairy calves. Twenty bull calves were assigned at birth to a milk level, fed via a teat: (1) *ad libitum* (ADL), or (2) a rate of 5 l/d, in two feedings (LIM). All calves were offered concentrate *ad libitum* during milk-feeding. Calves were weaned incrementally during wk 7 and then fed a pelleted diet *ad libitum* for 7 wks. Behavior was recorded from video for 3 d in each of wks 6, 8, and 14. Data was analyzed using a repeated measures mixed model. Calves fed ADL consumed more milk than LIM (in wk 6, 15.9 vs. 5.0 l/d, SE=0.8, P<0.001) in more meals/d (7.2 vs. 2.0, SE=0.8, P<0.001). All calves had similar sucking time/meal (6.0 min, P=0.8) and intake/meal (2.8 L, P=0.2). Including non-nutritive sucking bouts in the analysis revealed similar (P>0.6) bout frequency and sucking time/bout for all calves. During milk feeding, solid feed intake patterns differed; calves fed LIM consumed more than ADL (0.42 vs. 0.05 kg/d, SE=0.08, P<0.001), had longer meals (9.32 vs. 2.33 min/meal, SE=1.4, P<0.001), greater rates of intake (18.5 vs. 8.2 g/min, SE=4.0, P=0.02), and tended to have more meals/d (10.2 vs. 7.5, SE=1.4, P=0.07). Once weaned onto the novel pelleted diet, all calves had similar (P>0.2) intakes (1.6 kg/d), meal frequencies (13.6 meals/d), and average meal durations (15.4 min/meal), but calves previously fed LIM had greater rates of intake during meals (10.4 vs. 7.3 g/min, SE=1.5, P=0.04). All calves had similar meal patterns by wk 14; however, meal patterns changed over time (P<0.001). Meal frequency increased by 15.3%, meal duration decreased by 25.1%, and rate of intake increased by a factor of 3.3, resulting in an increase of 0.19 kg/meal. The results indicate that calves fed different milk levels develop similar meal patterns despite exhibiting unique milk and solid feed meal patterns early in life.

## Behavioural patterns of dairy heifers fed different diets

*Greter, Angela M.[1], Von Keyserlingk, Marina A. G.[2] and Devries, Trevor J.[1], [1]University of Guelph, Kemptville Campus, Animal and Poultry Science, 830 Prescott St, Kemptville, Ontario, K0G 1J0, Canada, [2]University of British Columbia, Animal Welfare Program, 2357 Main Mall, Vancouver, BC, V6T 1Z4, Canada; agreter@uoguelph.ca*

Little is known about how different types of diets affect the behaviour of intensively housed dairy heifers. The objective of this study was to describe the behaviour patterns of dairy heifers fed *ad libitum* in comparison to those fed at a restricted level (limit-fed). Twelve Holstein heifers (282.5±22.5 d of age; mean ± SD) were divided into 2 groups of 6 and exposed, in sequential replicates, to each of 2 dietary treatments using a crossover design with 7-d periods (2-d adaptation, 5-d data collection period). The treatment rations were: (1) high-concentrate TMR fed in a limited amount (2.0% of BW) with chopped wheat straw provided *ad libitum* (HCR+S), and (2) low-concentrate TMR fed *ad libitum* (LCR). Feeding, lying, and rumination behaviour were all recorded through use of automated monitoring systems. Data were analyzed in a mixed model including the fixed effects of treatment, period, and treatment×period, the random effects of replicate and heifer within replicate, and the residual error. Dry matter intake (DMI) was greater on the LCR treatment compared to the HCR+S treatment (7.3 vs. 5.3 kg/d, SE=0.2; P<0.001). Feeding time differed between treatments; while on the LCR, heifers spent more time feeding compared to when they were fed HCR-S (195.4 vs. 119.9 min/d, SE=8.3; P<0.001). We also noted no differences between treatments in feeding rate over the course of the day (0.05 kg/min; P=0.9). On the HCR+S treatment, prior to consuming their straw, heifers spent 79.9±15.7 min/d consuming 4.47±0.47 kg/d of TMR at a rate of 0.06±0.02 kg/min. Heifers spent an additional 40.0±23.9 min/d consuming 0.86±0.43 kg/d of straw at a rate of 0.04±0.01 kg/min during the rest of the day. On the HCR+S diet heifers spent less time lying than on the LCR (813.8 vs. 851.1 min/d, SE=13.4; P=0.007). Heifers on the HCR+S treatment also spent more time standing inactively (506.3 vs. 393.5 min/d, SE=14.9; P<0.001) but also more time ruminating (216.5 vs. 82.0 min/d, SE=33.7; P<0.005) compared to when they were fed the LCR. In conclusion, providing dairy heifers with a HCR+S not only reduced time spent feeding but increased inactive standing and decreased lying time, known risk factors for lameness.

## The effect of lameness to feeding behavior of dairy cow

*Tamminen, Petro[1], Korhonen, Arja[1], Häggman, Johanna[1], Jaakkola, Seija[1], Hänninen, Laura[2] and Pastell, Matti[1], [1]University of Helsinki, Department of Agricultural Sciences, P.O. Box 28, 00014 University of Helsinki, Finland, [2]University of Helsinki, Reserch centre for animal welfare, P.O. Pox 57, 00014 University of Helsinki, Finland; petro.tamminen@helsinki.fi*

Lame cows are reluctant to walk which may affect negatively to their feeding behaviour in loose houses. To study this we locomotion-scored 53 cows (1-4 parities) every second week for 8 weeks, and collected their daily feeding data. The cows were in loose house with automated milking system and had free access to silage from 22 troughs. They were fed concentrates, according to the stage of lactation, from two automatic feeders. The feeding troughs registered the visit times and durations automatically. The cows were assigned a locomotion score (LS) from 1 to 5. The cows with LS >= 3 were considered lame, and the LS >= 4 were considered to be severely lame. Scores 4 and five were combined for statistical analyze. The effect of lameness, parity and lactation stage on the mean daily feeding duration, feeding bout length and the incidence of visits were analyzed using repeated linear mixed effects models. The prevalence of lameness was 34%. There was a significant (P<0.001) interaction between LS and stage of lactation for the total feeding time. The lameness had no effect on the daily feeding duration at the early lactation, but decreased for the severely lame cows (LS >=4) in mid-lactation (152 min LS >= 4 vs 188 min LS=2 P<0.05 and 215 min LS=3 vs. 152 min LS=4 P<0.001). In the end-lactation the daily feeding duration decreased when the cow had locomotion score of >= 3 (248 min LS 2 vs. 217 min LS 3 P<0.001)). There was a significant (P<0.001) difference in feeding bout duration between sound and lame cows. The average daily feeding bout durations in minutes were 17.3 for LS=1, 16.4 for LS=2, 15.8 for LS=3 and 14.7 for LS:s 4/5. We found that the stage of lactation has a great impact on the feeding behavior of lame cows. The earlier the stage of lactation, the less the lameness effects on feeding behavior. It may be that the greater energy need at the early lactation may hinder the pain-related changes in the feeding behaviour. Towards the end-lactation and calving the cows also get heavier which may increase the lameness-related pain.

**How do different amounts of solid feed in the diet affect time spent performing abnormal oral behaviours in veal calves?**

*Webb, Laura[1], Bokkers, Eddie[1] and Van Reenen, Kees[2], [1]Wageningen University, Animal Production Systems Group, Animal Science, P.O. Box 338, 6700 AH Wageningen, Netherlands, [2]Wageningen University and Research Centre, Livestock Research, P.O. Box 65, 8200 AB Lelystad, Netherlands; laura.webb@wur.nl*

Veal calves fed mostly milk replacer and little solid feed display abnormal oral behaviours (AOB) (i.e. tongue playing/rolling, excessive oral manipulation of pen, and grazing other calves' coat). Indicative of chronic stress, AOB likely originate from a frustrated drive to chew and ruminate. This study investigated the effects of varying solid feed amounts on AOB in veal calves. Group-housed HF bull calves (N=48; 7.6±0.1 weeks; 54.7±0.3 kg) were fed four diets (A, B, C, D) including milk replacer and solid feed (50% concentrates, 25% maize silage, 25% straw of DM) in different amounts (0, 9, 18, and 27 g DM/kg$^{0.75}$/day). Twice a week, AOB were observed 1 h pre and post morning and evening feeding (i.e. 4 sessions per day), i.e. when AOB are most frequent. In each session, one calf per treatment was observed continuously for 10 min, thus 32 calves were observed per week. Chew/ruminate behaviours (CRB) and AOB were quantified throughout the day (06:30-19:00) once a week using scan sampling for 30 min every 2 h. The study lasted four months. Data were grouped by pen and month and analysed using a GLMM with fixed effects of month, treatment and month x treatment, and with a random pen effect. Calves in treatment D displayed less AOB than calves in A and C around feeding (%observed time: A=42±3, B=32±3, C=36±3, D=24±3; P=0.003) and AOB increased in study months 2, 3 and 4 compared with month 1 (1=16±2, 2=40±3, 3=38±3, 4=42±4; P<0.001). Throughout the day, no effect of treatment on AOB was found, but A calves tongue played more than calves in B and D, and C calves tongue played more than D calves (%scans: A=5±0.5, B=3±0.4, C=5±0.4, D=2±0.3; P=0.030). In month 1, C and D calves chewed more than A calves, and B calves chewed less than D calves (A=11±1, B=15±1, C=21±2, D=30±2; P<0.05). In month 2, D calves chewed more than A calves (A=10±1, D=18±2; P=0.023). We conclude that only the diet comprising of roughly six times the EU requirement for solid feed (i.e. D) provided a welfare benefit by reducing time spent performing AOB around feeding and tongue playing throughout the day, and by increasing CRB in months 1 and 2 relative to treatment A.

**Policy changes to enable sows to express behavioural needs in intensive housing conditions**

O' Connor, Cheryl and Cross, Nicki, Ministry of Agriculture and Forestry, Animal Welfare Directorate, P.O. Box 2526, Wellington, New Zealand; Cheryl.OConnor@maf.govt.nz

Intensive farming practices are increasingly coming under public scrutiny and methods being used to manage farm animals are becoming more important in the eye of the consumer. It is becoming more generally accepted that ensuring that production animals are not suffering physical discomfort is not enough, we must also allow them the opportunity to express their behavioural needs. This requirement to allow animals to express normal behaviour is reflected in the primary legislation in New Zealand, the Animal Welfare Act, 1999, and all animal welfare policy made under the Act. One farming industry that has recently come under intense scrutiny is the pork industry. Traditionally, many pregnant sows have been housed in crates to increase the rates of embryo implantation and to reduce social aggression and competition for resources. In recent years however, the extent of the behavioural restriction that this type of system places on sows has been increasingly recognised and a move away from the use of sow stalls has begun. Pregnant sows have been successfully housed and managed in group housing systems in a number of countries; significantly reducing the amount of time that they spend in close confinement and enabling them to express their behavioural needs. Group housing systems do, however, require increased amounts of labour and additional expertise on the part of the stockman to ensure that aggression and injury to sows are managed. Upon a recent revision of the Code of Welfare for Pigs (2010) the decision was made to phase out the use of sow stalls and restrict the use of farrowing crates in New Zealand. Development of animal welfare policy in New Zealand requires that, among other considerations, scientific knowledge and public opinion are taken into account and that a report outlining the significant differences of opinion and reasons for the recommendations made is also submitted when a code is recommended to the Minister of Agriculture. This paper will discuss the systematic, evidence-based approach that was used to encompass the range of social, ethical, economic and animal management factors that had to be taken into account when justifying the decision to change management practices in the pig industry in New Zealand.

## Determining the floor space requirement for group housed sows

*Rioja-Lang, Fiona C., Hayne, Stephanie M. and Gonyou, Harold W., Prairie Swine Centre, P.O. Box 21057, 21057 8th Street, Saskatoon, S7H 5N9, Canada; Fiona.lang@usask.ca*

With the announcements in 2007 by the largest producer/packers in both the USA and Canada that they will transition their production facilities to group housing for sows over the next 10 years, many producers are anticipating change. A frequently asked question is how much space do group housed sows require? The purpose of this study was to determine: the appropriate space allowance for group housed sows; how space allowance can influence posture and aggression; and if grouping sows with specific behavioral characteristics minimizes mixing aggression. It is thought that temperament could affect the ability of sows to compete within a group system. Sixteen groups of 8 sows were used. Groups were either uniform 'active' or 'passive' based on their response to two behavioral tests. Sows were confined to the loafing area for 23 hrs/day and returned to stalls for feeding. Panels were used to create space allowances of: 1.6, 2.0, 2.4, and 2.8 $m^2$/sow. Aggressive behaviors were observed live for 4 hrs at mixing. Photographs were taken for 72 hrs post mixing. Injuries were assessed and saliva samples were collected. The aggression, injury and cortisol data were analyzed using the Proc GLM procedure in SAS. The fixed effects included temperament, space and social group, and random effects of time period and replicate. Lying behavior was consistent throughout gestation. Over an average 24 hr period we observed 17% of sows standing, 2% sitting, 10% lying on sternum, 28% lying relaxed on sternum, and 43% lying laterally. Using calculations of the total area occupied by sows in these postures over 24 hrs it was calculated they would require 1.51 $m^2$/sow. This calculation is from a purely physical perspective and does not take into account the space required for social interactions or free space. The largest space requirement occurred between midnight and 8 am when the highest percentage of sows were lying laterally. When first grouped, sows showed a higher occurrence of injuries (P<0.001) and a greater number of fights (P<0.001) compared to 3 wks post-mixing. Grouping sows with different behavioral characteristics does appear to minimize aggression as 'passive' animals had lower injury scores and were involved in less fights, however this was not significant. There was no significant difference between injury score or number of fights with space allowance.

## Alleyway width in a free-access stall system influences gestating sow behavior and welfare

*Mack, Laurie[1], Eicher, Susan[2], Johnson, Anna[3], Lay, Jr., Donald[2], Richert, Brian[1] and Pajor, Edmond[4], [1]Purdue University, 125 S Russell St, W Lafayette, IN 47907, USA, [2]LBRU, USDA-ARS, 125 S Russell St, W Lafayette, IN 47907, USA, [3]Iowa State University, 2356F Kildee Hall, Ames, IA 50011, USA, [4]University of Calgary, 3280 Hospital Dr NW, Calgary, AB T2N 4N1, Canada; lmack@purdue.edu*

Free-access stalls allow sows to choose the protection of a stall or use of a shared alley. This study investigated the effect of alley width on behavior in gestating sows. At 35±8 d of gestation, 21 sows (N=126) were assigned to 1 of 3 pen treatments. Each pen contained 7 free-access stalls with 7 sows and a shared alley of 0.91, 2.13, or 3.05 m wide (1.86, 2.60, and 3.16 $m^2$ respectively). Sows remained in the pens until moving to farrowing stalls at approximately gestational d 111. Behavior was recorded once weekly for 24 h through the experiment and scored using 10-min scan samples. Data were analyzed in SAS using Proc Mixed with a post-hoc Tukey-Kramer adjustment. Sows with 0.91 m were observed more frequently in stalls ($P<0.05$) and less frequently in the alley ($P<0.01$) than either other treatment. As gestation progressed, sows used stalls less frequently ($P<0.001$) and alley more frequently ($P<0.0001$), but alley use increased least in 0.91 m pen ($P<0.01$). Sows with 0.91 m were observed more frequently partially in a stall than sows with 3.05 m ($P<0.05$) and sows with 2.13 m sows were intermediate. Stalls near the barn center were used more frequently than those near the outside wall ($P<0.05$) with middle stall use intermediate. Stall use by location showed no treatment differences. Lying was observed more frequently in sows with 0.91 m than those with 3.05 m ($P<0.05$), with 2.05 m intermediary. Standing decreased ($P<0.01$) and sitting tended to increase ($P<0.10$) as gestation advanced. Combined water and feed trough use tended to decrease during the first 5 wk on treatment and then increase until sows entered farrowing stalls ($P<0.10$). There were no differences by time or treatment for oronasal contact with pen walls or floor. Sows with 0.91 m were less frequently observed in social groups than sows with 2.13 m or 3.05 m ($P<0.05$) and the size of the groups was smaller ($P<0.05$). Social group size increased over time ($P<0.05$) with those in 3.05 m alley having the greatest increase ($P<0.01$). These results suggest that space limitations in the 0.91 m alley constrained the sows' behaviors.

## Action-reaction: using Markov analysis to elucidate social behavior when unacquainted sows are mixed

*Marchant-Forde, J.N.[1], Garner, J.P.[2], Lay Jr., D.C.[1] and Johnson, A.K.[3], [1]USDA-ARS-LBRU, W. Lafayette, IN 47907, USA, [2]Purdue Univ., W. Lafayette, IN 47907, USA, [3]Iowa State Univ., Ames, IA 50011, USA; marchant@purdue.edu*

Sows fight when mixed but there is little information on the detailed social behaviors performed by unacquainted sows at mixing. This project aimed to determine the sequence of behaviors leading up to aggression when unacquainted sows were mixed in pairs or in two established sub-groups of three, in different environments. Indoor experiments used York × Landrace sows. Expt. 1 used 11 sow pairs mixed into a 6.2 $m^2$ pen. Expt. 2 used 14 groups of 3 sows, with 2 groups mixed into a 19.2 $m^2$ pen. Outdoor experiments used Berkshire sows. Expt. 3 used 16 sow pairs mixed into a 5000 $m^2$ grassy paddock. Expt. 4 used 28 groups of 3 sows, with 2 groups mixed into a 5000 $m^2$ grassy paddock. In all experiments, behavior was recorded continuously for 60 min post-mixing. All-occurrences sampling was used to extract social interactions with each initiator behavior being followed by a recipient behavior, giving a sequence of reciprocated behaviors for each interaction. After extraction, the data were subject to a 1st order Markov transitional analysis to distinguish behavioral pairs that occurred more or less often than random. When mixed in pairs indoors, only 14.0% (random, n.s.) of withdrawal behaviors, such as retreat and head tilt, were followed by break – i.e. the interaction ending, whereas outdoors, the same behaviors were followed by break 47.0% (more, $P<0.05$) of times. Indoors, 32.3% (more, $P<0.05$) of bites were preceded by the recipient sow showing no response to the previous behavior, as opposed to 18.4% (random, n.s.) outdoors. With sows mixed in groups, 46.5% (more, $P<0.05$) of no response instances were followed by bites indoors, whereas, 30.4% (random, n.s.) were followed by bites outdoors. In the other direction, break was preceded by withdrawal 61.1% (more, $P<0.05$) of times outdoors but only 15.6% (more, $P<0.05$) of times indoors. Overall, the results indicate that, indoors, not responding to, or trying to avoid, the other sow did not necessarily diffuse the interaction, given the limited space, whereas outdoors, sows were able to use the space to effectively end aggression. The information derived from this study furthers our understanding of factors that may promote or ameliorate aggression in different mixing environments.

**Effects of enclosure size and complexity on captive African elephant activity patterns**
*Scott, Nancy L, Fripp, Deborah and Booth-Binczik, Susan D, Dallas Zoo, Conservation and Science, 650 South RL Thornton Fwy, Dallas, TX, 75203, USA; nancy.scott@dallaszoo.com*

The Dallas Zoo opened 2 new, complex elephant habitats in the Giants of the Savanna Exhibit in May 2010. A large, 1-ha savanna area was designed to encourage elephant activity with varied topography, deadfall, vertical snags, sand piles, boulders, a mud wallow, and filtered pools. A smaller, 0.2-ha close-up area has several of these elements as well as hanging hay nets. The zoo currently has 6 adult, female African elephants, in 2 social groups alternating between the 2 habitats. Two of these elephants had been previously housed in the zoo's old, less complex, 0.3 ha exhibit. To assess the effects of the new exhibits on these 2 elephants' behavior, their activity budgets were monitored before and after moving into the new area. We predicted the more complex habitats would increase foraging and walking time and decrease rates of stereotypic behavior (e.g., swaying). For comparison, activity budgets were also monitored for the 4 new elephants in the new area. Behavior was recorded at 1-min intervals for 15 min for each elephant on a randomized schedule during normal zoo operating hours (0900–1700). Data were analyzed using multi-factor ANOVA and post hoc Tukey's tests. For the 2 elephants that had lived in the zoo's old facility, stereotypic behavior decreased by 93.5% in the new, large habitat compared to the old yard ($P<0.05$). Foraging rates increased by 58.7% in the large habitat and 55.9% in the smaller, close-up exhibit compared to the old yard ($P<0.05$). For the other 4 elephants in the new spaces, rates of stereotypies were low in both the large (2.7%±0.6), and the small habitat (5.4%±1.2) and were most frequent when hay was absent. Foraging rates were high in both the large (50.1% ±1.6) and small exhibits (42.8% ±1.6). Walking rates were greater in the large habitat (11.8% ±0.6) than in the small habitat (9.3% ±1.2). Rates of these behaviors in each new habitat were similar between the original 2 elephants and the 4 new elephants except foraging rates were higher in the smaller exhibit for the original elephants ($P<0.05$). Overall, these results indicate that increasing the environmental complexity of enclosures increases the time elephants spend foraging and decreases the time spent involved in stereotypic behaviors. Additional data is being collected as the enclosures are modified further.

## Gender differences in stereotypical behavior can be predicted by gender differences in activity in okapi

*Fripp, Deborah[1], Watters, Jason[2], Bennett, Cynthia[3], Binczik, Gerald[1] and Petric, Ann[4], [1]Dallas Zoo, 650 S RL Thornton Fwy, Dallas, TX 75203, USA, [2]Brookfield Zoo, 3300 Golf Rd, Brookfield, IL 60513, USA, [3]Detroit Zoo, 8450 W 10 Mile Rd, Royal Oak, MI 48067, USA, [4]Okapi SSP, 2240 S 3rd Ave, North Riverside, IL 60546, USA; deborah.fripp@dallaszoo.com*

Abnormal repetitive behavior (ARB), sometimes called stereotypical behavior, is a significant welfare concern for animals in captivity. ARBs tend to increase in situations known to be suboptimal and are linked to other signs of poor welfare, including decreased reproductive success. In species comparisons, ARBs in captivity reflect activity in the wild, with wider ranging species expressing more locomotory ARBs than species with smaller home ranges. As males and females of some species have different behavioral profiles in the wild, including different home range sizes, gender can be expected to predispose individuals of those species to different ARBs. The home ranges of male okapis are 2-3 times the size of female okapis' home ranges. We hypothesized that based on their natural history, pacing should occur more in male okapis than females. To test this hypothesis, we performed a 3-year, multi-institutional study, including 53 okapi, 26 females and 27 males, at 18 institutions across the US. Each okapi was observed daily in a 15-min focal animal sample during five 90 d sessions that ran Jan-Mar or Jun-Aug. We recorded a wide group of activity budget behaviors on 1-min time points, as well as all occurrences of behaviors commonly seen as part of ARBs in okapi. Behavioral patterns were analyzed by 3-way ANOVAs, using gender, season, and climate (area of the US) as factors. As expected, male okapi were more locomotory, walking more (P=0.01) and pacing three times as much as females (P=0.001), while females foraged more (P<0.001). Interestingly, pacing and walking were not correlated ($r_{p}$_0.15, P=0.03). Oral stereotypies, such as object licking and tongue activity, were seen relatively equally in both males and females. These results indicate that in species such as okapi, the management needs of males and females should be considered separately. As with many ungulates, oral stereotypies are a concern for both genders. Locomotory stereotypies, on the other hand, appear to be primarily a concern for male okapis and may require us to reconsider the size and complexity of males' enclosures.

## Personality and stereotypy components in okapi

*Watters, Jason[1], Fripp, Deborah[2], Cynthia, Bennett[3], Binczik, Gerald[2] and Petric, Ann[4], [1]Chicago Zoological Society, 3300 Golf Road, Brookfield, IL 60513, USA, [2]Dallas Zoo, 650 S. RL Thornton Freeway, Dallas, TX 75203, USA, [3]Detroit Zoo, 8450 West 10 Mile Rd, Huntington Woods, MI 48070, USA, [4]retired, Okapi Species Survival Plan, Chicago, IL, USA; jason.watters@czs.org*

Animal personalities occur when individuals are consistent in their behavioral expressions across contexts or over time. Understanding how animals environments affect the expression of different personalities has potential for improving animal welfare and promoting the success of zoo missions. First steps to gaining this understanding should include (1) measuring animals' behavioral types and (2) determining whether alternative personalities differ in factors that reflect welfare or other forms of success. Here, we look specifically at the expression of stereotypy components to determine if individuals that act in consistently different ways in their normal daily routines also vary in their expression of this welfare measure. In a three-year multi-zoo study of okapi, we measured individuals' activity budgets for 90 days in both winter and summer. We also recorded the frequency of expression of components of potential stereotypic behavior. These components are the most basic elements of some abnormal behaviors. For example, while grooming itself is not abnormal, excessive grooming is. In okapi, potential stereotypies include locomotory, self-directed and oral behaviors, and behaviors directed at other animals such as allo-grooming. Using factor analysis on activity data we determined 5 personality components: communicative, appetitive, associative/exploratory, interactive, and active. We looked for location and sex effects on the expression of behavioral types and investigated whether certain types were more likely to express stereotypy or particular stereotypy components than other types. While we detected no location effect, in our sample, males are more communicative (urine mark, vocalize) and more active (play, manipulate objects) than females. Intriguingly, the personality of individuals relates to their expression of specific stereotypy components. For example, more active animals are more likely to pace, and other factors are associated in various ways with other stereotypy components. It may benefit managers to consider the behavioral types of individuals when planning population management and husbandry practices.

## Individuals interacting with environmental enrichment: a theoretical approach

*Franks, Becca[1], Reiss, Diana[1], Cole, Patricia[2], Friedrich, Volney[1], Thompson, Nicole[1] and Higgins, E. Tory[1], [1]Columbia University, 1190 Amsterdam Ave, New York, NY 10027, USA, [2]Vassar College, 124 Raymond Ave, Poughkeepsie, NY 12604, USA; beccafranks@gmail.com*

Research into individual differences in animal behavior is flourishing. We apply a novel motivational perspective from human social psychology, Regulatory Focus and Regulatory Fit, to the study of nonhuman primate behavior. Extending similar research conducted with laboratory rats, we tested whether a stable motivation to focus on gains over losses (a promotion-predominance) vs. losses over gains (a prevention-predominance) would (a) produce stable individual differences in the general behavior of zoo-housed primates and (b) predict how those individuals interact with enrichment objects. Observing a small group of cottontop tamarin (Saguinus oedipus) for 6 months, we characterized animals as having a promotion- or prevention-predominance with respect to several key behaviors: eating in the open, hiding, and time spent near the front of the exhibit. Using these characterizations, we predicted several distinct patterns of individual-object interaction. First, we predicted that compared to other individuals, a prevention-predominant individual would display more vigilant behavior when approaching enrichment objects. Second, that a promotion-predominant individual would approach enrichment objects faster than a prevention-predominant individual only when the object represents a potential gain. When the object represents a non-gain/potential loss, we predicted that a prevention-predominant individual would approach faster. We found support for our predictions, demonstrating that individual differences in motivational concerns can predict variability in how individuals interact with environmental enrichment. These results are discussed in relation to animal welfare science.

**Do impoverished environments induce boredom or apathy in mink?**

*Meagher, Rebecca, Diez Leon, Maria and Mason, Georgia, University of Guelph, Animal & Poultry Science, 50 Stone Road E., Bldg. 70, Guelph ON, N1G 2W1, Canada; rmeagher@uoguelph.ca*

Captive animals in impoverished environments are often considered 'bored'. In some contexts, particularly when individuals become profoundly inactive, they have also been called 'apathetic', 'lethargic' or other terms suggestive of depression. However, the validity of these terms has rarely been investigated, even though alleviating boredom is one suggested means by which environmental enrichment programs aim to improve welfare. Boredom in non-humans is hard to assess empirically, but as it is a negative state resulting from monotony or under-stimulation, we can predict that it would increase responsiveness to stimuli of any kind. Apathy (lack of interest), by contrast, would be manifest by decreased responsiveness to all stimuli; and depression-related anhedonia (loss of pleasure) would decrease responsiveness to typically rewarding stimuli, but not aversive ones. We tested the competing hypotheses that mink, a model carnivore, experience boredom or a depression-like apathy and/or anhedonia in standard fur-farm cages. We exposed 29 mink to a series of ten stimuli categorized a priori as aversive (e.g. an air puff or predator scent), rewarding (e.g. a moving brush to chase) or neutral (e.g. a plastic bottle or peppermint scent). In these 5-min tests, we assessed latency to make contact and attention (total time oriented to the stimulus) as indicators of responsiveness. Effects of long-term housing were tested using repeated measures GLMs controlling for sex and stimulus order. Non-enriched, standard-housed mink (NEE) made contact sooner ($F_{1,196}$= 18.48, P<0.0001), and attended to stimuli for longer ($F_{1,217}$= 41.12, P<0.0001) than mink housed with environmental enrichment (a double cage with running water and 'toys'; EE). The increased attention in NEE mink was only to aversive and neutral stimuli (treatment*type $F_{2,217}$=9.31, P=0.0001).However, when three food treats were offered in separate tests, NEE mink ate more of one type than did EE mink, suggesting they are more responsive to some rewards ($F_{2,53}$=4.77, P=0.01).Perhaps unsurprisingly given the lack of evidence of 'apathy', inactivity levels assessed via scans three weeks earlier did not predict responsiveness, within or across housing treatments (all P>0.05). Altogether, NEE mink show a heightened responsiveness to stimuli of all types, consistent with boredom.

## Effects of shade on feeding behaviour and feed intake of female goat kids

*Guevara, Nallely, Reyes, Manolo, Sánchez, Alejandra, Gamboa, Débora, De Luna, Belem, Galindo, Francisco and Alvarez, Lorenzo, Facultad de Medicina Veterinaria y Zootecnia, Universidad Nacional Autónoma de México, Producción Animal, Tequisquiapan, 07960, Mexico; alorenzo@unam.mx*

Solar radiation and high ambient temperatures negatively affect feeding time, performance, and animal welfare in several species. The provision of shade is a simple method that helps to minimize the negative effects. To determine whether shade influences feeding behaviour and feed intake in female goat kids, 40 dairy goat kids were used in two similar pens whose feeders were shaded (n=20) or unshaded (n=20) during 60 days. From May to July (Summer, 150 km North of Mexico City), behavioural data were collected by 10-min scan samples and for periods of 24 h. Both pens were shaded on the opposite side to the feeder. All goat kids were observed for their position inside the pen, and the number of times they were seen eating was recorded. When the concentrate was provided (13:30 h; concentrate test), the time was recorded until >50% of the animals stopped feeding. Refusals were collected and weighed daily to calculate the consumption. Ambient temperature and black globe temperature were daily recorded. Shade had a significant effect on the percentage of animals seen eating ($P<0.05$). A higher percentage of animals feeding was recorded in shaded feeders than unshaded ($P<0.05$). Feed refusal was higher in the unshaded feeder ($30\pm1.8\%$) than in shaded ($25\pm1.9$; $P=0.05$). The concentrate test duration was 26.6 min ($\pm1.3$) in shaded feeders and 16.1 min ($\pm1$) in unshaded ($P<0.05$). The concentrate test duration was negatively correlated to the ambient temperature in the unshaded animals ($r=-0.50$, $r2=0.25$; $P=0.02$), and it was not significant in the shaded ones ($r=-0.23$, $r2=0.05$; $P>0.05$). Results suggest that shade on feeders helps to ameliorate some negative effects of solar radiation increasing feeding time and feed intake in female goat kids. This could be of great interest to prevent performance and welfare negative affectations. DGAPA (UNAM), PAPIIT IN205810.

## Zoo-housed chimpanzees and gorillas are highly selective in their space use: implications for enclosure design, captive management and animal welfare

*Ross, Stephen R.[1] and Calcutt, Sarah J.[2], [1]Lincoln Park Zoo, Lester Fisher Center for the Study and Conservation of Apes, 2001 North Clark Street, Chicago, IL 60614, USA, [2]Emory University, 201 Dowman Drive, Atlanta, Georgia 30322 USA, USA; sross@lpzoo.org*

The relationship between physical environments and nonhuman primate behavior is critical to providing effective care and management in a range of settings. The physical features of the captive environment, including not only gross usable space but also environmental complexity, can have a significant influence on primate behavior and ultimately, animal welfare. But despite this connection, there remains relatively little data on how captive primates use the spaces provided to them. This might be particularly relevant in terms of great apes, for which resource-intensive, indoor–outdoor enclosures have become more prevalent in recent years. In this study, we examined where 23 great apes (chimpanzees and gorillas) positioned themselves within a modern zoo enclosure to determine not only how the apes utilized their space but also how access to outdoor areas affected their spatial selectivity. Using an electronic map interface, we recorded subjects' position every 20 seconds: both in terms of their location within a grid of 1 m × 1 m quadrants, and five vertical tiers. This resulted in 1896 hours of data (14840 data points per subject) collected over four years. We then characterized the degree to which apes spread themselves across all possible locations with the use of a Spread of Participation Index (SPI). We found that both species used relatively little of their available space: chimpanzees and gorillas spent half their time in only 3.2 and 1.5% of their usable three-dimensional space respectively. Chimpanzees utilized the outdoor space more than gorillas (t=7.01, P<0.001) but access to the outdoors did not affect the degree to which either species used their indoor areas. These results should be of use to both scientists and managers as they relate to enclosure design, captive management and animal welfare. The relatively narrow range of locations regularly accessed by animals in captive environments offers important insight into how animals perceive and value differential spaces provided to them.

## The effect of diet on undesirable behaviors in zoo gorillas

*Hoellein Less, Elena[1,2], Bergl, Richard [1,2], Ball, Ray[3], Kuhar, Christopher[1,2], Dennis, Pam [1,2], Raghanti, Mary Ann[4], Lavin, Shana[5], Wensvoort, Jaap[6] and Lukas, Kristen[7], [1]Cleveland Metroparks Zoo, 4200 Wildlife Way, Cleveland, OH 44109, USA, [2]Case Western Reserve University, 10900 Euclid Ave, Cleveland, OH 44106, USA, [3]North Carolina Zoological Garden, 4401 Zoo Parkway, Asheboro, NC 27205, USA, [4]Kent State University, 800 E Summit St., Kent, OH 44242, USA, [5]University of Florida, University of Florida, Gainesville, FL 32611, USA, [6]Toronto Zoo, 361A Old Finch Ave, Scarborough, ON 51B5K7, Canada, [7]Tampas Lowry Park Zoo, 1101 W Sligh Ave, Tampa, FL 33604, USA; eah@clevelandmetroparks.com*

In zoos, it can be difficult to provide diets that are analogous to natural diets. In the wild, western lowland gorillas travel long distances while foraging and spend large amounts of time feeding. In contrast, many gorillas in zoos consume a low volume, calorically dense diet which may contribute to observed low levels of feeding and foraging and overall activity. Typical captive gorilla diets may also lead to the formation of 'undesirable behaviors' or behaviors that do not occur in the wild or occur at a higher rate in zoos. To test the effect of diet on behavior, we implemented a biscuit-free diet at five institutions: North Carolina, Cleveland, Seattle, Toronto, and Columbus. We hypothesized the change would reduce undesirable behaviors including regurgitation and reingestion (R/R), decrease time spent inactive and increase time spent feeding. A repeated measures ANOVA revealed an elimination of R/R in 3 individuals and a significant reduction (by 43.6%) of the behavior (P=0.01) while coprophagy increased (P=0.00). This is the first scientific evidence of R/R elimination with a diet change in gorillas. Feeding significantly increased (P=0.01) and time spent inactive increased (P=0.02). However, both feeding and time spent inactive showed an institutional effect in that they increased at some institutions and not others. Hair plucking only occurred in Cleveland and was reduced (pre-diet: M=2.24, SE=1.68; post-diet: M=0.39, SE= 0.11) while ear covering only occurred in Columbus and increased (pre-diet: M=0.80, SE=0.47; post-diet: M=2.99, SE=1.37). The observed increase in coprophagy and ear covering requires further attention. This diet change has implications to animal management in that it ameliorates (and in some cases eliminates) certain behaviors considered to be undesirable in gorillas such as R/R and hair plucking.

## How animals win the genetic lottery: biasing birth sex ratio results in more grandchildren

*Thogerson, Collette[1], Brady, Colleen[1], Howard, Richard[1], Mason, Georgia[2], Pajor, Edmond[3], Vicino, Greg[4] and Garner, Joseph[1], [1]Purdue Universiy, 915 W. State St., W. Lafayette, IN 47906, USA, [2]University of Guelph, 50 Stone Road E, Guelph, ON N1G 2W1, Canada, [3]University of Calgary, 3280 Hospital Drive NW, Calgary, AB T2N 2Z6, Canada, [4]San Diego Zoo Global, P.O. Box 120551, San Diego, CA 92112, USA; collette@purdue.edu*

Population dynamics models and the physiology of mammalian sex determination predict that, on average, a parent should have equal numbers of sons and daughters. However, in theory, a parent can maximize fitness by biasing its birth sex ratio (BSR) in favor of offspring which will outperform peers. Sex Ratio Manipulation (SRM) theories propose cues which parents could use to alter their BSR, and in nature dams do bias BSR using such cues. All SRM theories share a fundamental prediction – grandparents who bias their BSR should produce more grandoffspring via the favored sex. However, despite many examples of biased BSRs, this prediction has never been directly tested in mammals. Using 100 years of breeding records from San Diego Zoo Global, we constructed a 3 generation pedigree containing 1627 granddams and 703 grandsires from 198 species. All analyses were GLMs and controlled for species, order, date of reproduction, number of F1 offspring, and selection of F1 animals for breeding. Dams and sires who bias their BSR had more grandoffspring in total ($P<0.0001$; $P<0.0108$) and the number of grandoffspring was mediated via the biased sex. The more male-biased a grandparent's BSR, the greater the number of grandoffspring via male F1 offspring (granddam, $P<0.0001$; grandsire, $P<0.0001$). The more female-biased a granddam's BSR, the greater the number of grandoffspring via female F1 offspring ($P=0.0272$), but not for grandsires ($P=0.9426$). Thus, biased BSR results in a greater number of F2 descendents and that success depends on the biased sex. These data confirm the ultimate reason why parents control BSR – parents who have cues to the future success of their F1 offspring and bias BSR accordingly have a clear F2 fitness advantage over those who cannot. Thus, SRM is a widespread and highly adaptive evolutionary strategy in mammals. However, in a captive population, this individually adaptive strategy may significantly impact long term survival of the species as parental control of BSR has the potential to accelerate genetic loss and risk of extinction.

**Reliability and validity of a subjective measure to record changes in animal behaviour over time**

Bishop, Joanna[1,2], Gee, Phil[1] and Melfi, Vicky[2], [1]University of Plymouth, Plymouth, PL4 8AA, United Kingdom, [2]Whitley Wildlife Conservation Trust, Paignton, TQ4 7EU, United Kingdom; joanna.bishop@plymouth.ac.uk

Studies which track changes in behaviour over time typically require long periods of observation by a trained researcher. However in zoos and aquariums there are often relatively large numbers of volunteers who could each invest smaller amounts of time in research, and pool data. This requires a methodology entailing little training that is both reliable between observers, and valid in terms of its measurement of behaviour. The methodological research reported here tested the reliability and validity of a scale termed 'busyness' which has been developed for use in long-term behavioural studies. Busyness is a subjective appraisal of how 'busy' captive animals are, on a scale of 1-5, considering the behaviour of all animals in an enclosure during a minute observation. To test reliability, 48 students rated the busyness of 20, 1-minute video clips of tiger behaviour. Results showed good Spearmans correlations between participants' scores ($P<0.001$) and results rarely differed by more than one busyness level for each video clip. To test validity, 39 students rated the busyness of 100 video clips of tigers, and this was compared to quantitative behavioural data recorded from the video using instantaneous scan and one-zero sampling of state behaviours every minute. Results showed positive Spearmans correlations between mean busyness scores and behaviours such as walking, trotting & running ($P<0.01$), and negative correlations with sleeping and inactive alert behaviours ($P<0.01$). Hence, busyness showed substantial agreement between observers, and correlated with behaviours related to locomotion, activity and alertness. In additional research designed to assess the application of the method in a zoo environment, busyness scores were collected throughout the day for a pair of tigers and showed increasing busyness levels as the feeding time approached. Further possible applications of this method will be discussed, such as in the study of predictable routines, or environmental enrichment.

## The positive reinforcement training effect: reduction of an animal's latency to respond to keepers' cues

*Ward, Samantha,[1] and Melfi, Vicky[2], [1]Moulton College, Animal Management and Welfare, Moulton, Northamptonshire, United Kingdom, [2]Whitley Wildlife Conservation Trust, Paignton Zoo Environmental Park, Devon, United Kingdom; sam.ward@moulton.ac.uk*

Positive reinforcement training (PRT) is increasingly adopted in zoos, to enable complex veterinary procedures to be undertaken without sedation or restraint and in some cases to reduce stereotypic behaviours. However, empirical studies to establish the efficacy and impact of PRT on keeper-animal relationships (KAR) are scarce and will be investigated in this study. Animals were classified as trained (T; undergo a systematic training regime), partially trained (PT; respond to keepers but no systematic training regime) or un-trained (UT; no systematic training regime). Eight black rhinoceros (2T, 2PT & 4UT), twelve Sulawesi macaques (4T, 4PT & 4UT) and eleven Chapman zebra (4T, 2PT & 5UT) were studied in six zoos across the UK and USA. Subtle cues and commands provided by keepers directed towards the animals were identified and the latency between these and the respective behavioural responses (cue-response) performed by the animals were recorded daily. A minimum of 5 cue-responses per keeper-animal dyad (n=76) were observed. Keepers also completed surveys about the animals' traits such as boldness and fearfulness, which were used to create ranked behavioural profiles. There were significant differences in cue-response latencies between species (ANCOVA: $F_{2,22}=25.017$, P<0.001), and social species (zebra and macaques) reacted significantly faster to cues than the solitary species (rhino) ($F_{2=}13.716$, P<0.001). There was no significant difference in the cue-response latencies according to behavioural traits (Wilcoxon: Z=-1.576, P>0.05), but all trained animals had shorter cue-response latencies than other animals ($F_{2,22}= 6.131$, P<0.01). Data suggests that group living animals are more receptive to PRT and it reduces the latency to react to cues in all animals, regardless of their behavioural profiles. Therefore the training can over-ride the individual's profile or tendency. Hosey (2008) suggested that low fear of humans in animals, is a necessary contributing factor to the development of a positive KAR. If we consider short cue-response latencies to be indicative of low fear, then we suggest that PRT potentially reduces animals' fear of humans thus can lead to positive KAR and thus increases animal welfare.

## Separating the stressors: a pilot study investigating the effect of pre-mixing calves on the behavior and performance of dairy calves in a novel environment

*Stanton, Amy L[1], Brookes, Raymond A[2], Gorden, Patrick J.[2], Leuschen, Bruce L.[2], Kelton, David F.[1], Parsons, Rebecca L.[2], Widowski, Tina M.[1] and Millman, Suzanne T.[2], [1]University of Guelph, 50 Stone Road W, Guelph, N1G 2W1, Canada, [2]Iowa State University, 1600 S 16th St., Ames, IA 50011, USA; astanton@uoguelph.ca*

The majority of milk-fed calves in North America are individually housed and the transition to group housing following weaning can be stressful, resulting in increased susceptibility to disease. The objective for this research was to determine the effect of separating the stressors of post-weaning mixing and movement to a novel environment. The hypothesis was that pre-mixing calves in the nursery would reduce the stress associated with these changes as demonstrated by increased feed intake and decreased activity during the post-movement period. Weaned calves were randomly assigned to either a pre-mixed or traditionally raised treatment. Pre-mixed calves (n=32) were grouped in the nursery barn by removing dividers between pens on Day 0 to form a large pen of 4 calves. Traditional calves (n=32) remained individually housed until Day 7. On Day 7, calves were moved to a new barn where they were grouped by treatment in pens of 4 calves each. Pre-mixed calves remained in their original social group. Calves were weighed on Days -1, 6, and 13. grain intake was measured at the group level on Day -3 through day 12. On Day -4 an accelerometer (IceTag®) was attached to the mid-metatarsal region. Daily averages for activity and calf starter intake were calculated for baseline (Day -1 and -2), mixing (Day 0-6) and post-movement (Day 7-13) periods. Associations between treatment, behavior, starter intake and average daily gain (ADG) were analyzed using linear mixed models. Calf starter intake was not significantly different by treatment during mixing or post-movement periods (P=0.34 and 0.55, respectively). Pre-mixed calves tended to gain more during the mixing period than the traditional calves (0.9±0.5 kg, P=0.10). There was no difference in weight gain during the post-movement period between treatments (P=0.73). During the post-movement period, pre-mixed calves tended to rest longer (P=0.09) and take (327±204) fewer steps per day than traditional calves (P=0.10) Pre-mixing calves tended to increase ADG premixing and reduce activity changes associated with movement to a novel environment.

**Dairy welfare in three housings systems in the upper Midwest**

*Lobeck, Karen, Endres, Marcia, Godden, Sandra and Fetrow, John, University of Minnesota, 1364 Eckles Avenue, St. Paul, MN 55108, USA; miendres@umn.edu*

The objective of this observational study was to describe dairy welfare in 3 housing systems: compost bedded pack (CB), low profile cross-ventilated freestall (CV), and naturally ventilated freestall (NV) barns. The study was conducted on 18 commercial dairy operations with herd sizes ranging from 75 to 1600 lactating cows. All CV and NV barns had sand freestalls and CB barns mainly used sawdust for bedding. Cows were visually scored for locomotion (LS), hock lesions (HL), body condition, and hygiene once every season of the year with approximately 90% of cows scored on each farm at each visit. Herd records were used to describe annual mortality rates (total animals that died divided by average herd size). Lameness prevalences (1=normal locomotion, 5=severely lame; LS $\geq$3=lame, LS $\geq$4=severely lame, respectively) were (LSMean, %$\pm$SE) 6.4$\pm$3.7, 14.1$\pm$2.2 and 17.7$\pm$2.2 in CB, CV, and NV barns, respectively. CB barns had lower lameness prevalence than NV barns (P=0.03). CV barns were not different from CB and NV barns. Severe lameness prevalences were 1.6$\pm$1.4, 2.2$\pm$1.0, and 3.1$\pm$1.0 for CB, CV, and NV barns, respectively with no differences among housing systems. Hock lesion prevalences (1=no lesion, 2=hair loss, 3=swollen hock; HL $\geq$2) were 7.8$\pm$5.2, 30.9$\pm$4.7, and 27.8$\pm$4.7 for CB, CV, and NV barns, respectively. The CV and NV barns had higher hock lesion prevalence than the CB barns (P=0.003 and P=0.012, respectively). Severe hock lesions (HL=3%) were 0.7$\pm$2.6, 7.4$\pm$1.9, and 7.8$\pm$1.9 in CB, CV, and NV barns, respectively. There was a trend for CB barns to have lower severe hock lesion prevalence than CV and NV barns (P=0.09 and P=0.07, respectively) with no differences between CV and NV barns. Hygiene scores (1=clean, 5=dirty) were 3.18$\pm$0.11, 2.83$\pm$0.08, and 2.77$\pm$0.08 for CB, CV, and NV barns, respectively. Body condition scores were 2.91$\pm$0.03, 2.97$\pm$0.03, and 2.96$\pm$0.02 for CB, CV, and NV barns, respectively. There were no differences among the housing systems for hygiene or body condition score. Mortality rates (%) were 5.1$\pm$1.0, 5.8$\pm$1.1, and 5.0$\pm$1.1 for CB, CV, and NV barns, respectively with no differences among the housing systems. In conclusion, CB barns provided a more welfare friendly environment than CV and NV barns based on lower lameness and hock lesion prevalences and no adverse associations with body condition, hygiene, or mortality rates.

## The effect of distance to pasture on dairy cow preference to be indoors or at pasture

*Charlton, Gemma[1,2], Rutter, S. Mark[1], East, Martyn[2] and Sinclair, Liam[1], [1]Harper Adams University College, Animal Science Research Centre, Newport, Shropshire, TF10 8NB, United Kingdom, [2]Reaseheath College, Agriculture Department, Nantwich, Cheshire, CW5 6DF, United Kingdom; gcharlton@harper-adams.ac.uk*

Several factors influence whether dairy cattle prefer to be indoors or at pasture, including weather conditions and milk yield, but it is unclear how the distance between the two locations influences preference. This study investigated whether pasture access of 60 m, 140 m or 260 m from the indoor housing would affect dairy cow preference to be at pasture. Twenty four Holstein-Friesian dairy cows were used during the study, which took place in the UK from May to July, 2010. There were four 18 day experimental periods, with eight cows in each period, which were further divided into two groups of four cows. Following a training period the cows were randomly allocated to walk 60 m, 140 m or 260 m to pasture, over three, four day measurement periods. A video camera was used to record time spent indoors and outdoors 24 h/day and behaviour observations (07:00 h to 22:00 h) took place 6 times during each period to record how the cows spent their time in each location. The video data showed that cows spent, on average 57.8% ($\pm$3.44) of their time outside (either at pasture or on the track). One sample t-tests revealed this was different to 0% (t=16.80; P<0.001), 50% (t=2.26; P=0.031) and 100% (t=-12.28; P<0.001). ANOVA of the percentage time spent outside revealed that distance did not influence night time pasture use (21:00 h to 04:30 h) ($F_{2,8}$=0.16, P=0.851; 81.0% vs. 81.0% vs. 76.7%, for 60 m vs. 140 m vs. 260 m, respectively). In contrast, during the day (07:00 h to 21:00 h; from behaviour observations) the cows spent more time at pasture when they had to walk 60 m ($F_{2,80}$=10.09, P<0.001) than when they had to walk 140 m or 260 m (45.3% vs. 27.4% vs. 21.2%, respectively). Neither the indoor temperature humidity index (THI) (62.1$\pm$0.62; $R^2$=0.0067, P=0.557) or the outdoor THI (59.6$\pm$0.64; $R^2$=0.0087, P=0.608) influenced time spent outside. The results indicated that cows had a partial preference for pasture, which was influenced by distance to pasture during the daytime, but not at night. The fact cows did not reduce pasture use with increasing distance at night, but did during the day, suggests access to pasture at night is more important to them than access during the day.

**Owner visitation: clinical effects on dogs hospitalized in an intensive care unit**

*Johnson, Rebecca A.[1], Mann, F. Anthony[2], Mc Kenney, Charlotte A.[3] and Mc Cune, Sandra[4], [1]University of Missouri, College of Veterinary Medicine, Research Center for Human Animal Interaction, Clydesdale Hall, Columbia, MO 65211, USA, [2]MU- CVM, Director of Small Animal Emergency and Critical Care Services, Clydesdale Hall, Columbia, MO, USA, [3]MU- CVM, ReCHAI, Clydesdale Hall, Columbia, MO, USA, [4]Waltham Centre for Pet Nutrition, Human-Companion Animal Bond Research Programme, Waltham-on-the-Wolds, Melton Mowbray, United Kingdom; rajohnson@missouri.edu*

Little attention has been directed to effects of animal owners visiting their hospitalized pets. Heart rate (HR) and mean arterial blood pressure (BPMAP) increases, and pain levels may be indicators of stress in hospitalized dogs. We identified effects of owner visitation on HR, BPMAP, and pain in dogs hospitalized in an intensive care unit. A one-group repeated measures pretest-post-test design was used. At four intervals during the owners' visit, the dogs' HR was determined by palpation and auscultation, BPMAP was measured, and a pain score assigned via a modified Glasgow Pain Scale. Sixteen owners (13 females, 3 males, Mean age=52 years) and their hospitalized dogs (Mean age=7.5 years) participated. The owners were allowed to visit as long as they wished with their dog. The observed visits lasted from 10–99 minutes, Mean 51.30 minutes. The dogs' HR increased from baseline (Mean=100 beats per minute) to 5 minutes after the visit began (Mean=110; P=0.0079), and increased again at 5 minutes before the owner left (Mean=112; P=0.0458). The dogs' HR then decreased 5 minutes after the visit ended (Mean=108; P=0.1552), but not below baseline. The dogs' BPMAP measurement levels increased steadily through the visits, though not significantly (Baseline Mean=115, 5 minutes into visit=122 {P=0.22}, 5 minutes before the owner left=124 {P=0.05}, and 5 minutes after the visit ended=126 {P=0.10}). Dogs' pain scores decreased from baseline to 5 minutes into the visit (Mean=2.20 to Mean=0.70; P=0.50). Data collection is ongoing. This research may give dog owners insight into whether or not visiting their hospitalized dog is advisable for the dog.

**Robot milking does not seem to affect whether or not cows feel secure among humans**
*Andreasen, Sine Norlander and Forkman, Björn, University of Copenhagen, Department of Large Animal Sciences, Groennegaardsvej 8, 1870 Frederiksberg C, Denmark; sinen@life.ku.dk*

Robot milking (AMS) is increasingly common. The use of robots has supposedly decreased the interaction between the dairy cows and the farmer. A possible negative consequence of AMS on animal welfare could be a worsened Human-Animal relationship, which could result in animals that are fearful of humans. In this study the avoidance distance for cows from 19 farms with an ordinary milking parlor (mean number of cows tested on each farm = 69 (range 56-87), in all 1303 cows) was compared to 12 farms with AMS (mean number of cows tested on each farm = 69 (range 58-80), in all 831 cows). The sample size recommended by the European Welfare Quality® protocol was used. The average number of cows on milking parlor farms was 180 cows and the average number of cows on farms with AMS was 169 cows. In all 31 farms with Danish Holstein-Frisian cattle were included in the study. The avoidance distance (ADF) of the cows was measured according to the Welfare Quality® protocol. The test started approximately 15 minutes after morning feeding and was carried out at the feeding table. The test person was unfamiliar to the cows. The test person started 2 meters away and walked slowly towards the cow with one hand lifted in a 45° angle. When the cow withdrew itself, the distance in centimeters from the hand to the cows muzzle was estimated. If the cow could be touched the distance was set at 0 cm. For each farm the average avoidance distance was calculated. The result showed no significant difference in the ADF between cows milked in ordinary parlors and cows milked in AMS (t-test; P=0.39, t-value = 0.87, N=31). To investigate if the number of animals had an effect on the ADF a correlation between the avoidance distance and the number of animals on the farms was calculated, this was however, not significant ($R_s$= 0.16, P=0.52). The results show that the postulated decreased interaction between cows and handlers when using AMS does not affect the Human-Animal relationship measured as avoidance distance. Earlier investigations have indicated that Human-Animal relationship may affect the milk yield of dairy cattle. This is not supported in the current study however, neither for the cows in AMS (ADF/milk yield – $r^2$=0.1 P=0.31) nor for the conventionally milked cows (ADF/milk yield – $r^2$=0.06, P=0.32).

**Relationship between amount of human contact and fear of humans in turkeys**

*Botheras, Naomi[1], Pempek, Jessica[1], Enigk, Drew[1] and Hemsworth, Paul[2], [1]The Ohio State University, Animal Sciences, Columbus OH 43210, USA, [2]Animal Welfare Science Centre, University of Melbourne, School of Land & Environment, Parkville VIC 3010, Australia; botheras.1@osu.edu*

Studies with broilers and layers show an association between amount and type of human interaction and bird fear of humans. This study investigated the relationship between the amount of human contact commercially-reared turkeys received and their fear of humans. Thirty barns, each housing 5800 male turkeys ($0.32 \text{ m}^2$/bird) were studied. At 4, 8 and 12 weeks of age, a behavioral test was conducted to assess fear of humans using the approach and avoidance responses of the birds to a stationary and moving unfamiliar human. A video camera was used to record the number of turkeys on-screen (close to the human) during the moving and stationary phases of the test. Stockpeople recorded for 1 wk prior to each visit the amount of time they spent in the barn daily. Blood samples were collected from 30 turkeys in each barn at 12 wks to determine heterophil:lymphocyte (H:L) ratios. Production data were obtained. Correlations between the average amount of time spent in the barn, the average number of turkeys close to the human during stationary and moving phases, H:L ratio and production were determined. Stockpeople spent $28\pm13$ min/d (mean$\pm$SD) in the barn when the turkeys were 4 wks old, increasing to $53\pm17$ and $60\pm23$ min/d at 8 and 12 wks, respectively. There were no significant correlations between time spent in the barn and the number of birds close to the human during stationary phases at 4, 8 or 12 wks (all $P>0.05$). Time spent in the barn at 4 and 12 wks was positively correlated with the number of birds close to the human during moving phases ($r=0.49$, $P<0.01$ and $r=0.46$, $P<0.05$, respectively). The number of birds close to the human at 4 wks was positively correlated with bird survival (stationary: $r=0.82$, $P<0.05$; moving: $r=0.67$, $P<0.1$). At 8 wks, the number of birds close to the human during the moving phase was negatively correlated with H:L ratio ($r=-0.50$, $P<0.05$). There was some evidence that increased human contact was associated with reduced fear of humans in turkeys. The quality or type of human interaction is also likely to be important. Increased turkey fear of humans was associated with increased H:L ratios and reduced bird survival, suggesting bird performance may be affected by fear, through stress.

## Characteristics of stockperson interactions with pigs in swine finishing barns

*Crawford, Sara[1], Moeller, Steven[1], Hemsworth, Paul[2], Croney, Candace[1], Botheras, Naomi[1] and Zerby, Henry[1], [1]Ohio State Univ, Fyffe Ct, Columbus, OH 43210, USA, [2]Univ. of Melbourne, Melbourne, VIC, 3010, Australia; crawford.586@osu.edu*

The study evaluated the relationship between stockperson daily work routines/actions with measures of behavior and stress. Finishing barns (32 sites; >1000 pigs) managed by 41 independent, paid stockpersons, were assessed on consecutive days during the daily barn check. Human behaviors, sounds made and time in the pens/aisles were recorded on both days. Stroll and saliva tests were collected on all pens; alternating pens with half of each test completed each day. A pig count, within the camera's field of view (144 x 85 cm), was recorded at 5 s intervals from the video, and averaged within pen. Saliva was collected from two random pigs using Salivettes presented by the experimenter. The first pig was sampled in the rear of the pen, the second at the aisle when pens held ≤35 pigs. A third pig, middle of pen, was sampled if >35 pigs per pen. Saliva samples were pooled within pen and concentrations quantified in duplicate. Stockperson data were summed across pens and days of observation and reported on a per pen basis. Stockperson was modeled as a random effect to analyze variation in time spent in the barn. Time spent during the barn check ranged from 5.8 to 128.8 s per pen with an average of 36.4 s in the study. Time required for saliva collection ranged from 41.8 to 83.8 s per sample (max 120 s) across all barns. Salivary cortisol concentrations varied (P<0.01) across stockperson (barn fully confounded) with an average of 1.326 ng/ml and a range from 0.609 to 1.957 ng/ml. An average of 2.5 pigs were observed within the camera view, ranging from 1.2 to 3.8 pigs across barns. The correlation between time spent per pen and the frequency of words spoken (r=0.71) and verbal sounds (whistles, hoots) (r=0.72) indicated that stockpersons using verbal cues spent more time in observation (P<0.01). Cortisol concentrations increased as the time to obtain a saliva sample increased (r=0.22). Cortisol concentrations and count of pigs were positively correlated (r=0.13). Stockpersons varied significantly with regard to time in pens, verbal sounds and words spoken. Implications on pig stress levels require additional elucidation.

## Animal abuse and cruelty: an evolutionary perspective

*Patterson-Kane, Emily[1] and Piper, Heather[2], [1]American Veterinary Medical Association, Animal Welfare Division, 1931 N Meacham Rd, Ste 11, Schaumburg IL, 60173, USA, [2]Manchester Metropolitan University, Education and Social Research Institute, 799 Wilmslow Road, Didsbury, Manchester, M20 2RR, United Kingdom; emilypattersonkane@gmail.com*

Many disciplines and professions address social issues such as animal abuse, and evolutionary biology is one complimentary strand of this endeavor. An evolutionary perspective promotes understanding of the biological factors that may contribute to the occurrence of animal abuse either through predisposition, repurposing or distortion of natural motivational systems, or the breakdown of normal function due to a biology/environment dissonance. This review of the literature surveys explanations offered for aggression shown towards animals, why some individuals are violent towards animals even when this is not socially acceptable, and finally, how social acceptability is established and enforced. Consideration is given to a range of suggested abuse motivators such as dirty play, predatory instincts and deliberate deviance—as well as positive protective factors such as biophilia and awareness of evolutionary continuity. A broad view of these evolutionary perspectives suggests that aggression and even some forms of violence have a biological function in acquiring food, competition, maintaining social harmony, and defending vulnerable community members. Socially unacceptable violent behavior may represent a misfiring of these behaviors, especially in individuals who have not matured and integrated into a healthy, harmonious family and community. It may also indicate an individual who has malfunctioned in response to environmental pressures, mental illness or personality disorder. By integrating evolutionary perspectives into animal cruelty prevention and response initiatives we can attempt to de-stigmatize participation in prevention, help-seeking and treatment programs and recognize the ongoing need to reconcile human species-specific needs with the demands of our modern habitats. An evolutionary perspective also draws further attention to the need to understand the specific factors causing a person to abuse animals as, like other forms of violence, there is not a single root cause, and so no single corrective measure will apply in all cases.

## Behavioural and physiological methods to evaluate fatigue in sheep following treadmill exercise

*Cockram, Michael[1], Murphy, Eimear [2], Ringrose, Siân [3], Wemelsfelder, Francoise [4], Miedema, Hanna [5] and Sandercock, Dale [5], [1]Atlantic Veterinary College, University of Prince Edward Island, Sir James Dunn Animal Welfare Centre, 550 University Avenue, Charlottetown, PEI, C1A 4P3, Canada, [2]Faculty of Veterinary Medicine, Yalelaan 7, 3584 CL Utrecht, Netherlands, [3]Scottish Agricultural College, Kings Buildings, West Mains Road, Edinburgh, EH9 3JG, United Kingdom, [4]Scottish Agricultural College, Sir Stephen Watson Building, Bush Estate, Penicuik, EH26 0PH, United Kingdom, [5]The University of Edinburgh, Royal (Dick) School of Veterinary Studies, Easter Bush Veterinary Centre, Roslin, Midlothian EH25 9RG, United Kingdom; mcockram@upei.ca*

Previous observational studies suggested either that sheep do not become markedly fatigued by long journeys or that previous methods did not identify fatigue. A range of methods were used to identify fatigue in 7 pairs of 13-month-old, female and castrated male, non-breeding sheep walked on a treadmill at 0.5 m/s for up to 5 h (treatment) or for two 10-minute periods (control). Median ambient temperature 14 °C (range 10 to 17 °C). One sheep only walked for 4.5 h, but all other treatment sheep walked for 5 h without apparent difficulty. A repeated measures mixed model (treatment, time, treatment×time) analysis was undertaken. In treatment sheep, there was a proportionate decrease in the median frequency of the electromyogram (-0.004±0.0988) that was significantly different from control sheep (0.109±0.0988) ($P<0.05$). Treatment sheep did not lie down sooner (median latency 1.3 h, $Q_1$ 1.1, $Q_3$ 1.8) or for longer after exercise (49% of 24 h post-exercise period ± 5.7) than controls (1.7 h, $Q_1$ 1.0, $Q_3$ 2.4 and 61%±5.7, respectively) ($P>0.05$). After exercise, there was no significant difference between the times taken by treatment (7.0 s) and control (6.7 s) sheep to obtain a food reward in a maze ($P>0.05$). Observers (naïve to the treatments), using free choice profiling, could not identify qualitative behavioural differences between treatment and controls. Two groups of terms were identified: agitated/active to still/calm, and tense to relaxed/calm, the term weary was used once. Although previous studies showed that vigorous exercise can fatigue sheep, there was little evidence that prolonged gentle walking fatigues sheep. In this study, the speed and gradient of the treadmill was not sufficient to consistently cause fatigue.

**Do differences in the motivation for and utilisation of environmental enrichment determine how effective it is at eliminating stereotypic behaviour in American mink?**

*Dallaire, Jamie and Mason, Georgia J., University of Guelph, Animal & Poultry Science, 50 Stone Road E, Building #70, Guelph, Ontario, N1G 2W1, Canada; jdallair@uoguelph.ca*

Environmental enrichment typically reduces the amount of time animals spend performing stereotypic behaviours (SB) like pacing. However, this effect is highly variable: some individuals completely cease performing SB, while others are hardly affected. In American mink, Mustela vison, we tested the hypothesis that environmental enrichment has the smallest impact on SB in individuals who only infrequently use enrichments and/or are weakly motivated to do so. The locomotor stereotypies of 17 adult female mink living in a non-enriched home cage were quantified before and after they were given free access to an extra enriched cage containing manipulable objects, climbing structures, and swimming water. Motivational strength was later assessed by using a progressively weighted push-door to measure the 'maximum price paid' (MPP) for entry to the enriched cage, corrected for MPP to access food. The amount of time spent performing SB decreased by about two thirds after mink were given enrichment ($F_{1,15}$=10.35, P=0.006), but neither elevated MPP for enrichment access ($F_{1,10}$=0.78, P=0.398), nor frequent use of enrichments ($F_{1,12}$<2.50, P>0.140), predicted the degree to which this behaviour was reduced. Therefore, our initial hypothesis was not supported. We did, however, find some surprising relationships. Mink who spent the most time performing SB before enrichment later spent the least time interacting with objects and water in the enriched cage ($F_{1,13}$=4.74, P=0.049). These mink instead showed the largest increases in inactivity following enrichment ($F_{1,13}$=18.11, P=0.001). We therefore tested the post hoc hypothesis that the tunnels leading to the enriched cage reduced SB by facilitating inactivity. In support of this hypothesis, mink with frequent pre-enrichment SB later spent most of their inactive time in these tunnels, while mink with negligible SB instead preferred to rest in their home cages ($F_{1,13}$=10.07, P=0.007). The tunnels were inaccessible to humans, suggesting that stereotypic mink might be most motivated to avoid handling: a hypothesis now requiring further test. Our findings caution against making a priori assumptions about how animals will make use of environmental enrichment.

## The effect of pasture availability on the preference of cattle for feedlot or pasture environments

*Lee, Caroline[1], Fisher, Andrew[2], Colditz, Ian[1], Lea, Jim [1] and Ferguson, Drewe[1], [1]CSIRO, Animal Welfare, FD McMaster Laboratory, Armidale 2350, Australia, [2]The University of Melbourne, Faculty of Veterinary Science, Melbourne, VIC 3010, Australia; caroline.lee@csiro.au*

Intensive feedlot finishing is perceived to affect welfare because cattle cannot perform normal behaviours evident in pasture environments, such as grazing. The objective of this study was to determine cattle preference for spending time at pasture (5 ha) or in a feedlot (25 x 10 m) under two pasture availabilities, good (3900 kg DM/ha) and poor (1900 kg DM/ha). Five groups of Angus steers (n=6 animals per group; 454±9.3 kg body weight) were tested in the good and poor pasture treatments (6 days per treatment). A commercial pelleted ration was available *ad libitum* in the feedlot. Electronic tag readers at the entrance and exit to the feedlot monitored animal movements between pasture and feedlot and the cattle were fitted with IceTags™ to measure time spent lying and standing. Time spent eating in the feedlot was recorded with video cameras. Data were analysed using a linear model in ASREML. There was no significant effect of pasture treatment on total time spent in the feedlot (good 6.0 h, poor 6.1 h; P=0.9) nor on percent time spent standing (both 4.5 h; P=0.82) or lying (good 2.7, poor 2.5 h; P=0.31) within the feedlot. There was a tendency (P=0.06) for cattle to spend more time at the feeder when offered poor pasture (1.37 h) than good pasture (1.23 h) but group feed intake in the feedlot did not differ (P=0.19) between the good and poor pasture treatments (57.8 and 63.2 kg/day, respectively). Cattle spent more time standing in the paddock when offered good pasture (8.1 h) than when offered poor pasture (7.3 h; P=0.02) but lying in the paddock did not differ between good and poor pasture (10 and 10.6 h, respectively; P=0.31). There was little feedlot activity at night between 2200 h and 500 h. Feedlot feeding periods peaked at the start of the day (600 to 800 h) with 2 smaller peaks at 1200 h and 1800 to 2000 h. In conclusion, cattle preference for a feedlot or pasture environment was not influenced by pasture availability. CSIRO acknowledges the funding provided by Meat & Livestock Australia and the Australian Government to support the research and development detailed in this publication.

## Mobile laying hens

*Gebhardt-Henrich, Sabine G. and Fröhlich, Ernst, Centre for proper housing of poultry and rabbits, Burgerweg 22, CH-3052 Zollikofen, Switzerland; sabine.gebhardt@bvet.admin.ch*

Free-range laying hens are frequently kept in large flocks. Since individual identification is difficult it is not known how laying hens move in non-cage housing systems. In particular, it is unknown if hens remain in a particular area of the hen house and may form subgroups where individuals could get to know each other. RFID (Radio Frequency Identification) tags were attached to 5% (100 to 900) 9 to 15 month old laying hens of 12 free-range flocks (2,000 to 18,000 hens) in the hen house at night. Hens were randomly selected in all parts of the house where they slept. We noted where in the hen house which tags were attached. During three weeks we monitored which popholes the hens used. Hen houses (26 to 80 m long and 6 to 13 m wide) were equipped with two or three parallel broadside aviary racks. Popholes with RFID antennas were located at one broad side of the house. 78% of the hens left the house during three weeks. The hens that had slept on the rack close to the popholes were just as likely to exit through the popholes as the hens sleeping on the distant rack (Generalized Linear Model, GEE, $chi^2$=1.65, 2 df, NS). Hens were more likely to use popholes close to the site of tagging than farther popholes. This effect was stronger in larger flocks (GEE, $chi^2$=7.83, 2 df, P=0.02) and decreased with the number of days after tagging (GEE, $chi^2$=5.6, 1 df, P=0.018). After four weeks, 19-45% of the hens had used every pophole. However, the larger the house the fewer hens were registered at every pophole ($r^2$=0.51, N=11, P=0.008). We conclude that most laying hens do not use the full length of a large hen house every day but are mobile enough to use the whole length of their accommodations over periods of weeks and may not form stable spatial subgroups. A width of the house of up to 13 m did not prevent hens from using an outdoor run.

## Open water provision for pekin ducks to increase natural behaviour requires an integrated approach

*Ruis, Marko and Van Krimpen, Marinus, Wageningen UR Livestock Research, P.O. Box 65, 8200 AB Lelystad, Netherlands; marko.ruis@wur.nl*

Relevant literature regarding duck welfare was reviewed, to investigate possible welfare improvements that might be applied in pekin duck husbandry. The study, initiated by organized Dutch duck farmers, revealed that a lack of open water is the most important welfare issue in Dutch pekin duck husbandry. Based on experimental studies, the following conclusions were drawn with regard to water provision: (1) Provision of water solely by nipples is inadequate for a good care of the body, and for cleaning of eyes and nostrils. Bell drinkers, water troughs, showers and freely accessible open water make it possible for the birds to take care of their bodies. (2) The behavioural need for open water also exists in a straw system in which the foraging need is already partly provided. (3) Data from the UK and Germany demonstrate that the absence of open water increased the risk of dirty eyes and stuffed up nostrils. (4) Ducks prefer bell drinkers over nipples, troughs over bell drinkers, and freely accessible open (swimming) water over troughs. The attractiveness of showers was very variable among different experiments. (5) For fulfilling the natural behaviors, a constant access to open water seems not to be necessary. More research is advisable, to determine the frequency and length of access. (6) Provision of open water enhances feed and water intake of ducks, compared to ducks that have only access to water by nipples. This may result in an increased feed intake, but also in an increased feed conversion ratio. (7) Water use in an open water system in which water is permanently supplemented, can be doubled compared to a nipple system. Much water is spilled. It is therefore recommended that open water will be provided above a slatted floor or in a covered outdoor area. (8) Compared to the water from nipples, bacteriological quality of open water is obviously reduced. The next step is to develop an appropriate open water system, meeting requirements of both duck and duck farmers. The levels of eye pollution and stuffing of the nostrils are good indicators of duck welfare. It is recommended to include these parameters in a welfare monitor for ducks, as this provides important information for the farmer to optimize the water system and related management.

## Does water resource type affect the behaviour of pekin ducks (*Anas platyrhynchos*)?

*O'driscoll, Keelin[1] and Broom, Donald[2], [1]Teagasc, Animal and Bioscience Research Department, Grange, Dunsany, Co. Meath, Ireland, [2]University of Cambridge, Department of Veterinary Medicine, Madingley Road, Cambridge, CB3 0ES, United Kingdom; dmb16@cam.ac.uk*

This study evaluated effects of four water resource (WR) treatments on water related behaviours of Pekin ducks during bouts of behaviour at the WR. Ducks (n=3200) were randomly assigned to one of four treatments at 20 days (d) post-hatch: a chicken (CH) or turkey bell (TU), trough (TR) or bath (BA). There were 8 replicate groups of 100 ducklings per treatment. Treatments represented increasing levels of access to water: the beak tip in CH, beak in TU, head and beak in TR, and whole body access in BA. Behaviour was video recorded on d34 and d40 between 10:00 and 22:00. Six bouts of behaviour at the WR were observed continuously in their entirety in each pen on each recording day (n=384 bouts in total). Data were analysed using the Mixed procedure of SAS. Treatment had no effect on behaviour bout duration. However, bouts tended to be longer at d40 (05:19 [mm:ss]) than d34 (05:01; P=0.07), and ducks spent more time standing at d40(04:31) than at d34(03:48; P<0.01). Treatment tended to have an effect on standing behaviour (P=0.07), with ducks in BA tending to stand longer than CH (P=0.1). There was no effect on overall or % time spent drinking. Treatment affected the duration of bathing behaviour, which increased with level of access to water (P<0.05). Specifically, treatment affected the number of head-dips (P<0.01), with ducks in TR performing more than CH (P<0.01) and tending to perform more than TU (P=0.08), and duck/dive behaviour (P<0.001), with TU performing fewer than BA (P<0.001) and TR (P<0.01), and CH fewer than BA (P<0.05). In general, ducks with access to WR that permitted a greater level of access to water performed more bathing related behaviours, even though duration of WR related activity was similar across all WR treatments.

## Physiological and behavioral response of crossbred zebu dairy cows submitted to different shade availability on tropical pasture

*Ferreira, Luiz C. B.[1], Machado Filho, L. Carlos P.[1], Hötzel, Maria J.[1], Alves, Andréa A.[2], Barcellos, Alexandre O.[2] and Labarrère, Juliana G.[3], [1]Lab. de Etologia Aplicada, Depto. de Zootecnia e Des. Rural, Universidade Federal de Santa Catarina, Rod. Admar Gonzaga 1346 – Itacorubi., 88034-001, Florianópolis, SC, Brazil, [2]Embrapa, Parque Estação Biológica – PqEB s/n°, 70770-901, Brasilia, DF, Brazil, [3]Universidade de Brasilia, Campus Universitário Darcy Ribeiro, 70910-900, Brasília, DF, Brazil; pinheiro@cca.ufsc.br*

Shade and water are major welfare constrains for grazing cattle. However, many believe that, because zebu cattle and their crosses are well adapted to the heat, they may dispense shade. This study was designed to evaluate the effect of different shade availability and shape on the physiological and behavioral response of crossbred zebu dairy cows on tropical pasture. Four groups of three lactating cows were tested in a 4×4 latin-square design, with periods of three days, in the following treatments: without shade (WS), single artificial shade (AS), bush (B) and scattered trees (ST). The behavioral variables were recorded in scans every 10 min, from 8:30 h to 15:40 h by visual direct observation. Milk production, water consumption, rectal temperature and respiratory rate were recorded daily for all cows during the experimental period. Data were statistically analyzed by ANOVA. No differences were found for the studied variables when cows were in treatments B and ST. However, compared to the treatments WS and AS, cows presented higher frequencies of grazing (P≤0.03), and of shade use (P≤0.001), as well as lower rectal temperatures and respiratory rates (P≤0.005). Water intake (P≤0.002), ruminating (P≤0.003) and lying (P≤0.03) time were higher in B and ST than in WS, whereas cows showed intermediate values for these variables when in AS treatment. When in WS, cows showed greater time on other behaviors, presented the highest frequency of idling, lowest water consumption and greatest respiratory rate of all other treatments. It is concluded that the presence of abundant shade in pasture may prevent heat stress, indicating better welfare for crossbred zebu dairy cows raised on pasture in tropical areas.

### Changes is dairy cattle behaviour as a result of therapeutic hoof block application

*Higginson, Janet H.[1], Shearer, Jan K.[2], Kelton, David F.[1], Gorden, Pat[2], Cramer, Gerard[1], De Passille, Anne Marie B.[3] and Millman, Suzanne T.[2], [1]University of Guelph, 50 Stone Rd E, Guelph, ON N1G 2W1, Canada, [2]Iowa State University, College of Veterinary Medicine, Ames, IA 50011, USA, [3]Agriculture and Agri-Food Canada, 6947 #7 Highway, P.O. Box 1000, Agassiz, BC V0M 1A0, Canada; jhiggins@uoguelph.ca*

Lameness is one of the primary welfare concerns in the dairy industry. Although behavioural observation has been used for the detection of lameness, it has not been utilized for the study of efficacy of treatments such as the application of therapeutic hoof blocks for sole ulcers. Additionally, therapeutic hoof blocks may be underutilized due to concerns about their impact on behaviour and ability to compete for resources. Therefore, the objective of this study was to examine the effects of application of therapeutic hoof blocks on the behaviour of non-lame dairy cows. Wooden hoof blocks, 2.2 cm thick, attached with Bovi-Bond (Bovi-Bond, Netherlands) were randomly assigned to the left and right medial hind claws of 10 out of 20 sound Holstein cows housed in the same freestall pen and were observed for a total of 28 days. A subset of these cows was fitted with IceTag3Ds (IceRobotics, UK), which were affixed to both hind legs of 4 blocked and 4 control cows. Multivariable mixed modeling with repeated measures for cow was used to determine behavioural differences between the blocked and control animals. Behavioural changes were expected to occur shortly after block application; therefore Days 1 and 2 post-block application were compared to the day prior to application (pre-block). There were no significant differences in the number of steps taken between the two periods for either blocked or unblocked cows (P>0.05). On average, cows took 2574.3±241.5) steps per day pre-block and 2414.2±133.2) steps per day post-block (P=0.30). An average of 11.7±1.6) and 11.4±0.9) lying bouts per day were performed pre- and post-block, respectively (P=0.69). There was a trend for increased lying duration in the post-block period (pre-block = 58.1±5.2), post-block = 69.6±4.8) minutes per bout, P=0.07), but this was observed in both blocked and unblocked cows. There appear to be no significant changes in the activity or lying behaviour of dairy cattle during the 2 days following application of a block to a single hind claw.

**Effect of different environmental conditions in loose housing system on claw health in Finnish dairy cattle**

*Häggman, Johanna and Juga, Jarmo, University of Helsinki, Department of Agricultural Sciences, P.O. Box 28, FIN-00014 University of Helsinki, Finland; johanna.haggman@helsinki.fi*

In dairy farming, mobility is the most important prerequisite for the smooth operation of activities in a loose housing system where claw disorders are recognized as a major welfare problem. The aim of this study was to investigate how different environmental conditions, i.e. surface in stalls, feeding system and access to outdoors, affect the prevalence of claw disorders. The data used were collected by hoof trimmers in 2005–2009 and consisted of 8,349 Ayrshire and 4,406 Holstein cows in 306 loose housed herds. Eight different claw disorders, namely sole haemorrhage, interdigital dermatitis, sole ulcer, white-line disease, heel horn erosion, corkscrew claw, chronic laminitis and digital dermatitis, were combined to one binomial claw health trait. The data were analyzed using an R statistical software package. A logistic generalized linear model with hoof-trimmer and farm (within hoof-trimmer) as random effects were fitted to dataset. We found that Holstein cows had a 1.11 times higher risk of getting claw disorders compared with Ayrshire cows. Cows in parity 3 and ≥4 were more likely (OR=1.56 and OR=2.64, P<0.001) to get claw disorders than cows in parity 1. There were less claw disorders in barns that had corridors with free stalls (OR=0.89, P<0.001), deep litter with free stalls (OR=0.67, P<0.05) and deep litter without free stalls (OR=0.39, P<0.001), compared with farms that had slatted floor and free stalls. Farms with a flat rate feeding system had 2.32 times more claw disorders than farms that adjusted feeding according to yield. Cows that were in pasture in summer and had a possibility to go out in winter had less claw disorders than cows that were always inside (OR=1.12, P<0.01) or cows that had all year access to outdoors but were not in pasture (OR=1.34, P<0.001). Cows with straw as a bedding material had significantly more claw disorders in all bed surfaces (P<0.001) compared with cows that had shavings/sawdust or peat as a bedding material. To conclude, the results highlight the benefits of feeding adjustment according to yield, use of pasture in summer time, avoidance of slatted floors with free stalls system as well as the use of straw as a bedding material.

**Using behavior and physiology to assess the welfare of a rat model of multiple sclerosis**

*Hickman, Debra and Swan, Melissa, Indiana University, Laboratory Animal Resource Center, 975 W. Walnut St (IB008), 46202 Indianapolis, Indiana, USA; hickmand@iupui.edu*

A commonly used rodent model of multiple sclerosis involves the injection of Lewis rats with myelin basic protein (MBP) emulsified into complete Freund adjuvant (CFA). The inflammation at the immunization site (which can range from a footpad to a subcutaneous injection depending upon laboratory) and ascending paresis are potentially distressful for the animals. This study aimed to quantify the pain and distress experienced by animals used on these studies through the use of passive behavioral and physiologic assessments. Female Lewis rats were immunized at different injection sites with CFA. We hypothesized that stress-induced changes in behavior and physiologic parameters would be higher in rats immunized in the footpads as compared to those immunized subcutaneously at the tail base and non-immunized controls, but that all sites would be equivalent in inducing the intended disease. Telemetry transmitters were surgically implanted in the abdominal aortas of 54 Lewis rats to record body temperature, blood pressure, and activity. The rats were also digitally recorded for blinded ethological assessment. Nine rats were randomly allocated to 1 of 6 different immunization treatment groups: (1) hind footpads with MBP/CFA; (2) front footpads with MBP/CFA; (3) tail base with MBP/CFA; (4) hind footpads with hen egg lysoenzyme (HEL) in CFA; (5) hind footpads with saline; and (6) no manipulation. Blood pressure, temperature, and ethological profile over the course of the disease progress and histopathology and ability to induce EAE were compared between groups. Analysis of the physiologic results suggested elevations in blood pressure between groups 1 through 4 as compared to groups 5 and 6. The ethological data suggested that rats immunized in the front footpads spent more time engaged in behaviors consistent with distress. Microscopic evaluation of the heart and intestines did not reveal significant fibrosis of inflammation in any treatment group. Evaluation of the animal model (defined as time to onset of disease and positive lymphocyte proliferation assay) showed no significant differences between groups 1 through 3. The results of this study validate the use of subcutaneous, non-footpad injections, as a refinement to these studies.

**Handling of painful procedures in dairy calf management in Santa Catarina state, Brazil**

*Cardoso Costa, João H., Balcão, Lucas F., Darós, Rolnei R., Bertoli, Franciele and Hotzel, Maria J., LETA – Laboratório de Etologia Aplicada, Universidade Federal de Santa Catarina, Rod. Admar Gonzaga, 1346 – Itacorubi, 88034-001, Florianópolis, SC, Brazil; juaohcc@gmail.com*

Public concern is growing for the welfare of farm animals, particularly regarding procedures that cause pain or distress. A survey on the management of dairy cattle was carried out during the summers of 2009 and 2010, in the state of Santa Catarina, Brazil, with an interview of 120 dairy farmers. Here we report data on three common painful procedures applied to calves: dehorning, castration and extra teat removal. Only 5% of producers do not dehorn their calves; 92% dehorn their own calves; 3% hire veterinarians to carry out the procedure. The hot iron method, used by 74% of producers, is the most common practice used to dehorn calves. Seventeen percent use a hot electric iron, 4% use acid paste, and 1% perform scoop amputation. Although we found a wide variation among farms, dehorning is done at a median age of 5.1 months, and only 7% of producers dehorn their calves before 45 days. Most farmers raise one or more male calves per year; from those, 18% do not castrate them, 74% castrate their own calves and 8% hire veterinarians; median age is 5.2 months and only 17% of producers castrate their calves before 60 days. Surgical castration is used by 87% of the farmers, while 11% use emasculator and 7% use rubber rings. Whereas 75% of producers do not remove extra teats from their calves, 27% amputate without cauterization, 17% amputate and cauterize, and 7% use rubber rings. Importantly, although 96% of dairy producers stated that cows can feel pain, only one producer used methods of pain control during the dehorning procedures and none during teat removal or castration. Overall, we found that farmers perform these procedures at a late age, using methods which are painful for the animals, and do not use anesthetics. In order to attempt to change this situation, in a follow-up study we will investigate attitudes amongst these producers towards pain in animals, their intentions to change their practices, and whether they receive adequate information regarding these procedures.

## Strain variations in behavioral traits under heat stress in laying hens

*Felver-Gant, Jason[1], Mack, Laurie[1], Dennis, Rachel[2] and Cheng, Heng-wei[2], [1]Purdue University, Animal Science, Purdue Univeristy, 47907, USA, [2]Live Stock Behavior Unit, USDA-ARS, 125 South Russell Street, West Lafayette, IN, 47907, USA; hwcheng@purdue.edu*

High temperature (i.e., heat stress, HS) is a critical environmental factor affecting chicken welfare by decreasing birds' nutrition utilization, growth rate, and reproduction, and increasing mortality. This study examined whether hens selected based on one characteristic may affect their response to various stressors, such as HS. Ninety 28-week-old White Leghorns from two strains were used: DeKalb XL (DXL, 48 hens), a line of hens individually selected for high productivity, and KGB (kind gentle bird, 42 hens), a line of hens selected for high group productivity and survivability. The hens were randomly paired by line at 2 hens per cage, providing 658 $cm^2$ floor space per hen, and assigned to heat (H) or control (C) treatment for 14 days (mean: C=24.3 °C, H=32.6 °C, Humidity=30+5%; n=12). Behavioral data was recorded at day 1, 2, 6, 11, and 13 using 10 min scan sampling for 2 periods of time at 2 h for each, started at 2 h after lights on and 2 h before lights off, respectively. Data were analyzed using the mixed model procedure of the SAS program. Compared to the C hens, the H hens spent more time drinking and resting, displayed more wing-opening behavior, and less time sitting ($P<0.05$). Strain differences were also seen across the treatments. Compared to H-DXL hens, H-KGB hens were more active on day 1 ($P<0.05$), 11 and 13 ($0.05<P<0.1$); and tended to drink and eat more on both day 1 and 13 ($0.05<P<0.1$). H-KGB hens also exhibited more panting behavior than H-DXL hens ($P<0.05$). The results indicate that, KGB hens selected for high group productivity and survivability may have great capability to adapt HS. The method used for selecting the KGB line could be used by poultry producers to select chickens for increasing egg production, at the same time, improving bird welfare.

### Daily heart rate patterns of dairy cows in intensive farming conditions

*Speroni, Marisanna and Federici, Claudia, Agricultural Research Council (CRA), Fodder and Dairy Productions Research Centre (FLC), via Porcellasco 7, 26100 Cremona, Italy; marisanna.speroni@entecra.it*

Behavioral response to stressing situations is under the nervous and hormonal control; heart rate (HR) has been widely used as indicator of stress in dairy cows since it is at crossing of many physiological pathways that allow animals to monitor and control external environment. However, the diurnal pattern of HR in intensive farming is not very well known while the interest on biological rhythms and their adaptive functions is increasing. Seventeen lactating cows, from an herd of 70, were monitored for HR and behavior over a period of 24 h. The herd was milked twice a day in a double-8 herringbone parlor and housed in a free stall barn with slatted floors and cubicles. A total mixed ration was fed *ad libitum* once daily (8 h). Heart rate was recorded over 5 s intervals using a Polar monitor system consisting of two electrodes, an emitter and a watch-like receiver fitted on an elastic belt placed around the cow's chest, just behind the front legs. Individual data were summarized by hour. A mixed model for repeated measures was used to estimate the least squares means for each hour. Mean HR (bpm) resulted significantly ($P<0.0001$) affected by the hour of the day; it increased when cows were more active (eating, moving or milked) and reached the maximum value (95.7 ±1.4 bpm) at 20 h; then, progressively, mean HR decreased until 4 h (76.5 ±1.4 bpm). The reduction of HR during the night time can be easily explained by a general reduction in activity of the animals and the reduced environmental stimulation by human activities inside the barn. The cardiovascular changes that characterize the transition from activity to resting (decreased blood pressure, decreased peripheral resistance, decreased sympathetic tone, etc.) determined mean HR reduction during this period. Between 4 h and 6 h cows were collected and milked an this event interrupted the progressive decreasing of mean HR; mean HR decreased slightly more when cows came back the barn after milking, so that, unexpectedly, it reached the minimum (75.9±1.6 bpm) in the post-milking (8 h), instead of during the night resting. A possible explanation, supported by literature, could be the reduced metabolic load due to the fastening. From these preliminary results we conclude that diurnal HR in farming conditions is highly affected by feeding routine

**Euthanasia practice in Canadian animal shelters**

*Caffrey, Niamh, Mounchili, Aboubakar, Mcconkey, Sandra and Cockram, Michael, University of Prince Edward Island, Sir James Dunn Animal Welfare Centre, Atlantic Veterinary College, 550 University Avenue, C1A 4P3 Charlottetown, Canada; ncaffrey@upei.ca*

The objective of this study was to determine the methods of euthanasia used in Canadian animal shelters and identify any potential animal welfare concerns. Questionnaires on methods of euthanasia used were sent to 196 Canadian animal shelters yielding 67 responses. Nineteen percent of dogs and 40% of cats that entered a shelter were euthanized. The need for access by non-veterinarians to controlled drugs was an issue raised by 5 respondents. The services of a veterinarian were used for euthanasia in 82% of establishments. Sodium pentobarbital injection (a controlled drug) was the only method of euthanasia used by 61 and 53% of establishments euthanizing dogs and cats, respectively. Pre-medication was used by 58 and 48% of establishments that used sodium pentobarbital to euthanize dogs and cats, respectively. Injection of T-61 (a non-controlled drug in Canada) was the only method of euthanasia used by 23 and 35% of establishments euthanizing dogs and cats, respectively. All of these establishments used pre-medication, but the percentage of establishments that only used the intravenous route for administration of T-61 in dogs and cats was 45 and 7%, respectively, indicating that T-61 is not always used according to the manufacturer's guidelines. T-61 is a combination analgesic, anaesthetic and curariform drug. Euthanasia results from central nervous system depression, hypoxia and circulatory collapse. Its use is controversial due to concerns as to whether the curariform action causes respiratory paralysis before the animal losses consciousness. Following the administration of a pre-medication drug and T-61, vocalisation and twitching was reported by 8 and 11 respondents respectively. This method was rated as 3 'okay', on a scale of 1-5 in relation to the level of distress caused to the animal. The use of sodium pentobarbital was considered to be 5, 'optimal practice'. 'Causing no undue stress to animals' was recorded as best practice by 18 respondents. Use of pre-medications (12 respondents), having trained and competent staff (16 respondents), and having a dedicated euthanasia room (14 respondents) were also best practice opinions. Further research into the use of T-61 is required.

### Shelter dog behavior improvement: dog walking as enrichment

*Mc Kenney, Charlotte A.[1], Johnson, Rebecca A.[2] and Mc Cune, Sandra A.[3], [1]University of Missouri, College of Veterinary Medicine, Research Center for Human-Animal Interaction, Clydesdale Hall, Columbia, MO 65211, USA, [2]University of Missouri, College of Veterinary Medicine & Sinclair School of Nursing, Research Center for Human Animal Interaction, Clydesdale Hall, Columbia, MO 65211, USA, [3]Waltham Centre for Pet Nutrition, Human-Companion Animal Bond Research Programme, Waltham-on-the-Wolds, Melton Mowbray, United Kingdom; mckenneyc@missouri.edu*

Several million dogs are euthanized in animal shelters annually after multiple relinquishment reasons (Scarlett, 2002; Salman, 1998; New, 2000 & Kass, 2001). Gains in pet adoptions are happening via shelter enrichment programs. We hypothesized that shelter dogs participating in a daily dog walking program involving elderly citizens, would have better behavior, higher adoption rates, and decreased euthanasia rates than dogs in a control group not in the walking program. All participant dogs were pre-qualified for walking through the standard shelter behavioral assessment for adoption. The dogs, at least one year of age were matched with a control dog for size (small, medium and large). The experimental group walked with an older adult five days a week. The control group of dogs did not walk. Pre-test and daily behavior scores were assigned. The length of time each dog spent in the shelter was recorded as were adoption, move to foster care, release to a breed rescue group or euthanasia outcomes. There were 84 dog pairs. Outcomes for the experimental (walking) group: adoption n=58, to foster/ rescue n=13, euthanized n=7. For the control group: adoption n=26, to foster/rescue n=28, and euthanized n=20. A chi-square test showed that the experimental group had significantly more adoptions (P<0.0001) and fewer euthanasias (P=0.0063) than the control group. The control group had significantly more dogs that went to breed rescue networks (P=0.00071) than did the experimental group. The control group had a higher total behavior score (exhibited more negative behavior). The Wilcoxon rank sum test was used to compare the experimental and control groups in terms of total behavior scores. Dogs in the experimental group had significantly better behavior than dogs in the control group (P=<0.0001). The dog walking program was associated with desired dog behavior outcomes, better adoption rates and lower euthanasia rates.

## A new behavioural test for kittens before adoption

*Onodera, Nodoka, Mori, Yoshihisa and Kakuma, Yoshie, Teikyo University of Science, Department of Animal Sciences, 2-2-1 Senjyusakuragi, Adachi-ku, Tokyo, 120-0045, Japan; g061001@st.ntu.ac.jp*

There is an increased demand for promoting adoption of relinquished domestic cats, especially kittens, at animal shelters. Although it is important for the shelter to pass information on the kitten's temperament and behavioural style to new owners for the sake of successful adoption, few standardized behavioural testing has been established and widely used for kittens. There have been several studies which examined personalities of adult cats in laboratories or owned cats based on behavioural observation, but only a few studies were carried out with kittens. In this study, we aimed to develop a simple behavioural test which can be used to evaluate behavioural characters of kittens before adoption and easily applicable at shelters. The test consisted of eight items to examine the responses of kittens to novel place, toys, and an unfamiliar person (observer), and the number and latency for each response was recorded. Twenty-three kittens between 1-5 months of age were observed before adoption at a shelter. Only about half of the kittens came out of a cage within 3 minutes and mean (±SD) latency to come out was 61.0±40.9 sfor 11 kittens when an observer was there and the latency was shorter without an observer. The mean latency to approach a novel toy was 4.3±6.2 s for 9 kittens and to play was 4.9±4.3 s for 10 kittens. The latency to play when the observer moved her hand was 5.2±6.0 s for 6 kittens. Fifteen kittens approached the fingertip presented by the observer and the latency was 2.3±2.1 s. The latency to escape when held in the observer's arms was 10.0±6.6 s for 11 kittens and the number of attempts to escape ranged between 0 and 8. Principal component analysis extracted three elements in kittens' responses. They were labeled as 'Fearfulness', 'Withdrawal' and 'Going-my-way', respectively. Previous studies based on behavioural observation in laboratories suggested that cats could be roughly divided into three personality types such as active/aggressiveness, confident/easy-going, and timid/nervous. Our study also showed three underlying factors in kittens' characteristics using a simple behavioural test which can distinguish different characters and be carried out in a small room with a table. This would enable animal shelters and veterinary practices to employ this kind of behavioural testing.

## Relaxing effect of four types of aromatic odors in dogs

*Kuwahara, Yukari[1], Horii, Takayuki[2], Uetake, Katsuji[1], Iida, Yutaka[3] and Tanaka, Toshio[1], [1]Azabu University, 1-17-71 Fuchinobe, Chuo-ku, Sagamihara, 252-5201, Japan, [2]Yamazaki Gakuen University, 2-3-10 Shoutou, Shibuya-ku, Tokyo, 150-0046, Japan, [3]Yokohama Pet Communite College, 2-15-19 Shinyokohama, Kouhoku-ku, Yokohama, 222-0033, Japan; ma1007@azabu-u.ac.jp*

Recently, many aroma oils have been launched in Japan for both human being and pet animals due to the expectation of their relaxing effect. This study explored the effect of four types of aromatic odors (chamomile, peppermint, rosemary and lavender) on behavior and physiology of 12 naive dogs caged in two experiment institutions. The dogs were simultaneously exposed to each type of aromatic odor in the order described above after the evening feeding, through the diffusion of its essential oil, for 30 minutes a day for 5 days, with an interval of 2 days between odors. Five days of no odor exposure were set as a control. The dogs' behavior was observed every day, for 90 minutes including the exposure time. Saliva samples were collected on days 1, 3 and 5 to determine their cortisol levels. Dunnett's test was applied to compare with the control. Proportion of time spent standing significantly (P<0.05) decreased in three aromatic odors (chamomile: 5.9%, peppermint: 6.6% and rosemary: 4.7%) than the control (20.2%), and lying tended to increase instead (all odors: P=0.07). Dogs spent more time sleeping significantly (P<0.05) in rosemary (54.7%) than the control (22.8%). Frequency of gazing significantly (all odors: P<0.01) decreased with all aromatic odors (chamomile: 14.0%, peppermint: 11.9%, rosemary: 6.0% and lavender: 8.5%) than the control (34.2%). Frequency of moving round in the cage also significantly (all odors: P<0.05) decreased with all odors (chamomile: 0.5%, peppermint: 0.4%, rosemary: 0.2% and lavender: 0.3%) than the control (1.7%). Saliva cortisol level significantly (P<0.01) decreased in lavender (194.4 pg/ml) in comparison with the control (420.2 pg/ml). These results indicate that four types of aromatic odors used in this study have some positive effects, and particularly rosemary and lavender appear beneficial in their relaxing effect in dogs.

## The effects of space allowance and exercise for greyhounds on welfare

*Jongman, Ellen[1], Hemsworth, Paul[2] and Borg, Samantha[1], [1]DPI Victoria, Animal Welfare Science Centre, 600 Sneydes Road, 3030 Werribee, Australia, [2]University of Melbourne, Animal Welfare Science Centre, School of Land and Environment, 3010, Australia; ellen.jongman@dpi.vic.gov.au*

To investigate adequate floor space allowance for greyhounds in individual kennels, the effects of space and exercise outside the kennel on the behaviour, stress physiology and injuries of adult greyhounds were evaluated. Thirty-six greyhounds (aged between 12 and 17 months) housed in wire mesh kennels were studied in three time replicates in a 2×2 factorial experiment in which two main effects of floor space (3.0 vs 10.0 $m^2$) and exercise (none vs 20 min. daily) were examined over 6 weeks. General behaviour and activity were recorded by video during weeks 1 and 5. Saliva samples were collected during week 6 and analysed for basal cortisol concentrations. At the end of week 6 a blood sample was collected following an ACTH injection and injuries were assessed. Videos were analysed for time budgets of behaviour (lying, standing, walking and sitting), abnormal behaviour and the area of the kennel in which the behaviour occurred. The data were analysed using a multi-strata analysis of variance with treatment effects of exercise, age, floor space and interactions between exercise and each of the other 3 effects. Exercise had no significant effect on behaviour, physiology and injuries. Overall there were no significant effects of kennel size on behaviour and physiology, other than more time spent in the front of the large kennels ($P<0.05$). There was a tendency ($P<0.1$) for the younger dogs in the large kennel to spend less time at the back, less time lying down and more time walking. Similarly, younger dogs in the large kennels were less likely to lie down at the back of the kennel than dogs in the small kennels ($P<0.05$). From this study it is concluded that housing adult greyhounds in kennels with a floor area of 3.0 $m^2$ does not impose any greater welfare risk than housing in kennels with 10.0 $m^2$, regardless of exercise.

**Behavioral assessment in dogs during animal-assisted interventions (MTI)**

*Glenk, Lisa Maria[1,2], Stetina, Birgit Ursula[3], Kepplinger, Berthold [1,4] and Baran, Halina[1,2], [1]Karl Landsteiner Research Institute for Pain Treatment and Neurorehabilitation, Neurochemical Laboratory, LKM Mauer-Amstetten, 3362 Mauer/Amstetten, Austria, [2]University of Veterinary Medicine, Physiology, Veterinärplatz 1, 1220 Wien, Austria, [3]Workgroup Counseling Psychology, Department of Psychology, Berchtoldgasse 1, 1220 Wien, Austria, [4]Neuropsychiatric Hospital Mauer, Neurology, LKM Mauer-Amstetten, 3362 Mauer/Amstetten, Austria; lisa.molecular@gmail.com*

Animal welfare assessment through behavioral observation has become an increasingly important issue over the past years. Animal-assisted interventions are founded on beneficial effects of human-animal interaction on human psychological and physiological health. Despite the perception of high standards in animal-assisted interventions, undoubtedly the incorporated animal´s behavior is affected by interaction with clients. At present only few approaches to assess the welfare implications of these interaction effects on behavior exist. The dog-assisted group training MTI (multiprofessional animal-assisted intervention) is a carefully evaluated and goal-directed program that aims to improve human social and emotional competences. Interaction behaviors displayed by animal handlers and patients towards the dog during MTI include verbal contact, praising, tactile contact, gesturing, obedience commands, treat reward and playing with the dog. With the aim of documenting therapeutic dogs´ behavioral repertoires and activity budgets, we constructed an ethogram to cover welfare-related behaviors. Seven healthy dogs of different sex, age and breed were video-taped during 10-12 consecutive sessions that were carried out weekly in different institutions (inpatient drug withdrawal, prison, school). Gestures were evaluated using the Observer software package in relation to their frequency of occurrence during the 50 minute observation period. Behavioral taxonomy was chosen in accordance with earlier studies and includes general activity, yawning, licking the nose, paw lifting, turning the head, turning away, sniffing the ground, displacement activity, body shake and panting. In the present research, effects of human-animal interaction and the level of client familiarity on single behavioral parameters in therapeutic dogs were investigated. Preliminary results of the study will be presented.

## Behavioral and physiological evaluation of welfare in shelter dogs in two different forms of confinement

*Dalla Villa, Paolo[1], Barnard, Shanis[1], Di Fede, Elisa[1], Podaliri, Michele[1], Siracusa, Carlo[2] and Serpell, James A.[2], [1]Istituto G.Caporale, via campo boario,1, 64100 Teramo, Italy, [2]University of Pennsylvania, School of veterinary medicine, 3900 Delancey street, 19104-6010 Philadelphia, USA; p.dallavilla@izs.it*

A great part of the dog population in Western countries spends most of its life in kennels (e.g. working and shelter dogs). The typical kennel environment presents several stressful factors for a dog, and poor housing conditions can negatively affect the animal's welfare. In Italy, the National Law (281/1991) forbids the euthanasia of shelter dogs if not dangerous or seriously suffering; this leads inevitably to overcrowded facilities where welfare becomes a major issue. Previous studies have shown that social isolation decreases dogs' welfare, however, group housing is often not a viable solution. In this research project, the effect of two different forms of social housing were compared: group versus pair housing. Behavioral and saliva cortisol parameters were used as indicators of dogs' welfare. In the first housing condition, 17 subjects were housed in groups of 5-8 animals, in 4 outdoor enclosures (36 m$^2$). In the second housing condition, 8 experimental subjects were transferred in pairs (one male and one female) to smaller enclosures (6 m$^2$) while the remaining dogs were left in the outdoor enclosure as controls. Behavioral data and saliva cortisol samples were collected during 3 consecutive days in both conditions. A mixed linear model with subjects and housing as random effects, and their interaction as fixed effect was carried out. Saliva cortisol concentration decreased significantly (P=0.003) in both experimental and control dogs indicating that this parameter varied independently of housing conditions. Behavioral analysis showed that group housed dogs were significantly more active (T=3.82, P=0.002), they did more visual (T=3.49, P=0.003) and olfactory (T=2.42, P=0.03) exploration of the environment, they barked more (T=36, P=0.008) and sleep was interrupted more frequently (T=0, P=0.01) compared to pair housed dogs. Clinical data of the subjects collected throughout the study revealed no variation in the dogs' health condition. Advantages, disadvantages and the effects on dogs' welfare of both forms of confinement will be detailed in the discussion.

## Reactions of mother-bonded vs. artificial rearing in dairy calves to isolation and confrontation with an unfamiliar conspecific in a new environment

*Wagner, Kathrin[1], Barth, Kerstin[2], Hillmann, Edna[3] and Waiblinger, Susanne[1], [1]Institute of Animal Husbandry and Welfare, Vetmeduni, Veterinärplatz 1, 1210 Vienna, Austria, [2]vTI, Federal-Research Institute of Rural Areas, Forestry and Fisheries, Trenthorst 32, 23847 Westerau, Germany, [3]Animal Behaviour, Health and Welfare Group, ETH, Universitätsstraße 2, 8092 Zurich, Switzerland; kathrin.wagner@vetmeduni.ac.at*

The aim of this study was to investigate the effects of mother-bonded (M) vs. artificial (A) rearing in the first 12 weeks of life on the behaviour and stress reaction of dairy calves in challenging situations. M had unrestricted contact to their mother, the cow herd and the calf group. A were fed milk up to 16 kg/day, housed in a group of calves and had no access to the mother or cows. At the age of 43 days, they were tested in an isolation-test for 15 min (M and A both n=16). At an age of 90 days calves were tested in a social confrontation-test with an unfamiliar calf in a test-arena for 20 min (M n=13; A n=14). Data were analyzed using ANOVA with treatment, gender, breed as fixed factors and weight as covariate in both tests. In the confrontation test, longer transport before test, presence of cow next to test arena, partner with experience were additional factors. In isolation M showed more escape behaviour (M 3.97±0.82 (mean±SE); A 0.71±0.89 events/15 min, P<0.05), tended to be more vigilant (events/15 min, M 14.70±1.16; A 11.45±1.26, P<0.1) and to sniff more often (M 17.42±0.98; A 14.82±1.06 events/15 min, P<0.1) than A. M and A did not differ in heart rate. During confrontation M vocalized more than A (M 23.83±3.06; A 13.17±3.29 events/20 min, P<0.05), but both vocalized less in the presence of a cow and more if they were transported before. The percentage of received aggressive interactions of all aggressive interactions tended to be lower in M (M 35.58%±8.63; A 58.08%±8.77, P<0.1), furthermore M tended to initiate more aggressive behaviour than A (M 6.45±1.41; A 2.55±1.41 events/20 min, P<0.1), presence of a cow was associated with more initiated aggression in A and M. Heart rate tended to be lower in M than A (M 113.42±4.85; A 125.93±4.38 beats/min, P<0.1). The results suggest that mother-bonded calves show higher motivation to rejoin their herd and/or mother, show a more active coping style in both situations, higher social competence and they are less stressed by new social encounters.

## Sleep in dairy cows recorded with a non-invasive EEG technique

*Ternman, Emma[1], Hanninen, Laura [2,3], Pastell, Matti [2,4], Agenas, Sigrid[1] and Nielsen, Per Peetz[1],*
*[1]Swedish University of Agricultural Sciences, Animal Nutrition and Management, Kungsangen Research Centre, SE 753 23 Uppsala, Sweden, [2]University of Helsinki, Research Centre for Animal Welfare, Koetilantie 7 (P.O. Box 57), 00014 University of Helsinki, Finland, [3]University of Helsinki, Production Animal Medicine, Koetilantie 7 (P.O. Box 57), 00014 University of Helsinki, Finland, [4]University of Helsinki, Department of Agricultural Sciences, P.O Box 28, FI-00014 University of Helsinki, Finland; Emma.Ternman@slu.se*

The aim of this study was to develop a non-invasive technique for identifying different vigilance states in dairy cows. Sleep is a fundamental function and it is known that sleep deprivation both increases energy requirements and impairs immune defence and it is therefore possible that lack of sleep may contribute to animal welfare problems in dairy herds. Sleep is often estimated by behavioural observations or recorded on restrained animals with invasive EEG techniques. The latter might stress the animals and hence alter the sleep duration and distribution. A total of eight dairy cows were included in the study; five dry, of which three were of the Swedish Red breed and two of the Ayrshire breed, and three lactating dairy cows, all of the Ayrshire breed. Recording sessions were performed on one cow at a time and lasted until sleep-like rest had been observed. The cows were kept in single pens three hours before and during recording sessions. Before each session, the cows were equipped with surface attached electrodes measuring brain activity (EEG), eye movements (electrooculography EOG), and neck muscle activity (electromyography EMG) to record vigilance states. The recordings resulted in a total of 33 hours and 54 minutes of analyzable data (range from 1 h 44 min to 6 hrs 11 min per recording session). Data was scored manually for vigilance states, and the scoring was supported by behavioural registrations from direct observations. Rapid eye movement (REM) sleep and alert wakefulness shared similar features of desynchronized waves with varying high and low frequency and could be separated on account of the EMG data. Non-rapid eye movement (NREM) sleep displayed low frequency waves, sometimes with slow wave activity. The recorded data showed that it is possible to distinguish between different vigilance states in dairy cows using non-invasive EEG-technique but not by behaviour registrations alone.

## Effects of sloped standing surfaces on cattle behavior and muscle physiology

*Rajapaksha, Eranda and Tucker, Cassandra, University of Califronia, Davis, Animal Science, Room 1403, Meyer Hall, One Shields avenue, 95616, USA; earajapaksha@ucdavis.edu*

On dairy farms, flooring is often sloped to facilitate drainage. Sloped surfaces have been identified as a risk for lameness in cows, but relatively little is known about how this feature of flooring affects dairy cattle behavior. We evaluated the effect of slope on restless behavior, skeletal muscle activity, and latency to lie down after 90 min of standing. Sixteen Holstein cows stood on floors with 0%, 3%, 6%, and 9% slope for 90 min/treatment before milking in a cross-over design, with 24 h between each testing session. The order of exposure to treatments was balanced across the experiment. Electromyograms (EMG) were used to evaluate the activity of middle glutealand biceps femoris muscles. Contractions were recorded before, after and during exposure to each slope. Median power frequency (MPF) and median amplitude (MA) values were used for analysis. We also measured restless behavior, or the number of steps taken, and the latency to lie down after the test sessions. General linear models were used to compare both treatments and muscle groups. We predicted that restless behavior, muscle fatigue (as measured by MPF and MA) and latency to lie down after testing would increase with the slope of the standing surface. We found, however, no significant differences in muscle function, restless behavior or latency to lie down associated with slope. Myoelectrical activity (MPF) was greater for the middle gluteal muscle ($91.4\pm1.69$ Hz) compared to the biceps femoris muscle ($77.7\pm1.24$ Hz, $P<0.001$) indicating the middle gluteal muscle was more active under these test conditions. The number of steps increased over the 90-min of standing ($P<0.001$, from 4.5 to 6.5 steps/min in the first and last 15 min, respectively), regardless of the slope. Although restless behavior and muscle function did not change with slope in this context, this work is the first to use EMG to assess skeletal muscle activity in cattle. We suggest that this technology, along with restless behavior, may be useful in assessing muscle function, and perhaps fatigue, in more strenuous situations, such as prolonged standing associated with transport.

## Goats might experience motion sickness during road transportation

*Aoyama, Masato, Motegi, Takumi, Kaneta, Hiroki and Sugita, Shoei, Utsunomiya University, Department of Animal Science, 350 Minemachi, Utsunomiya-city, 321-8505, Japan; aoyamam@cc.utsunomiya-u.ac.jp*

Introduction: Road transportation can be a severe stress for domestic ruminants, but it is unclear whether they experience motion sickness because ruminants do not vomit. This study examined whether goats might experience motion sickness during road transportation.In experiment (Exp) 1, we observed the behavioral changes induced by the administration of cisplatin (CP), which causes emesis and vomiting in humans and dogs, in goats. In Exp 2, we examined the effects of the administration of diphenhydramine (DH), which prevents motion sickness in humans, on behaviours in goats during road transportation. Materials and Methods: In Exp 1, five adult Shiba goats were used. They were intravenously (IV) administrated with CP, and the behavior of each goat was videotaped. Each goat was administrated with the same volume of saline on another day for the control session. In Exp 2, eleven adult Shiba goats were used. They were intramuscularly (IM) administrated with DH 15 min before the start of 60 min road transportation, and the behavior of each goat was videotaped during the transportation. Each goat was administrated with the same volume of saline and transported on another day for the control session. Results and Discussion: In Exp 1, 133 min (45-189 min) after the administration of CP, all goats showed a specific postural shape for 238 min (55-410 min) consisting of a lowered head and little movement (even after human approach). These shapes were not observed in control session. It was possible that these behavioural changes were caused by the induction of a feeling of emesis. In Exp 2,there were no remarkable differences in behavior among control and DH session during 15 min pre-transportation period.During road transportation, some goats adopted a similar shape to that observed in Exp 1, but its duration in the DH session (18.4±4.39 min) was significantly shorter than that in the control session (24.8 ±4.81 min) ($P<0.05$, Wilcoxon's test). In addition, the attempts to escape from human approach during transportation in DH session (2.73±0.47 times) were significantly higher than those in controls (1.73±0.56 times) ($P<0.05$). These results provide some circumstantial evidence that is consistent with the suggestion that goats can experience motion sickness during road transportation.

**The reliability of rumination data recorded by a commercial rumination monitor**

*Rutter, Steven Mark, Brizuela, Carole and Charlton, Gemma, Harper Adams University College, Animals Department, Newport, Shropshire, TF2 8SR, United Kingdom; smrutter@harper-adams. ac.uk*

On-farm recording of rumination behaviour has the potential to contribute towards oestrus detection as well as in monitoring health and welfare status, and commercial rumination monitors have been developed with these aims. These devices also have the potential to be used for scientific studies if they provide a reliable method of recording rumination. The reliability of a commercial on-farm rumination monitor (Fabdec Heatime Vocal) (H) was evaluated by comparing it with an automatic jaw movement recorder (IGER Behaviour Recorder) (I). Concurrent 24 hr H and I recordings from three lactating Holstein-Friesian dairy cows collected on two separate occasions were analysed. Data were analysed as minutes ruminating per two hour period (the output format of H), and I data were taken as definitive, as the validation of I has been published. Two measures of rumination by H and I were compared: the total time spent ruminating over a 22 or 24 hr period, and the correlation between minutes ruminating per two-hour period. H consistently underestimated total time spent ruminating. For cow 1, H underestimated total time spent ruminating time by 11% and 9% on the first and second occasions. For cow 2, the underestimates were 4% and 3%, and they were 37% and 30% for cow 3. For cow 1, there was a very strong correlation in minutes ruminating per 2 hrs between I and H, with correlation coefficients (r) of +0.90 and +0.96 (n=11) on the first and second occasion respectively. For cow 2, there was a moderate correlation (r=+0.57 and +0.54, n=11). For cow 3, there was a very weak correlation on the first occasion (r=+0.15, n=11), although there was a strong correlation on the second occasion (r=+0.88, n=11). The slopes of all the lines fitted between I and H rumination data were less than 1.0, indicating that H tended to over-estimate ruminating in periods where there was little ruminating, and underestimated ruminating in periods where there was a lot. Although the consistencies in some aspects of the data recorded with cows 1 and 2 show the potential of H in scientific studies, the inconsistencies observed with cow 3 indicate the need for further research. In conclusion, although showing potential, further research is needed before the Fabdec Heatime collars can be used to reliably record rumination as part of scientific research projects.

## Sampling freqeuncy and duration for behavioural analysis and effectiveness of electronic tracking

*Maia, Ana Paula A.[1], Green, Angela R.[2], Sales, G. Tatiana[2], Moura, Daniella J.[1], Borges, Giselle[3] and Gates, Richard S.[2], [1]UNICAMP, Agricultural Engineering College, Campinas, SP, Brazil, [2]Univ of Illinois, Agricultural & Biological Engineering Dept, Urbana, IL, USA, [3]USP-ESALQ, Biosystems Engineering Dept, Piracicaba, SP, Brazil; sales1@illinois.edu*

Sampling frequency and duration for behavioral data collection is critical, so that data are collected efficiently but adequately to capture behavioral profile. No standard approach for behavior sampling was found in the literature. Reducing the time to collect behavioral data while maintaining the integrity of results can increase the scope and potential impact of behavioral studies. Additionally, the inclusion of simple electronic monitoring may be a valuable tool for improving behavioral data collection. The aim of this study was to determine optimal behavioral sampling for a study with laying hens and assess accuracy of IR sensors for tracking hen movements. Subsamples for this analysis were taken from four hens in different replicates of the full experiment. Video images were recorded for one hen housed in two wire mesh cages connected by acrylic tunnels, with a feeder in one cage, a drinker in the other, and companions in adjacent cages. Behavior was analyzed in 1 min segments over 24 h. Behavioral time budget included: eating ($26\pm5\%$), drinking ($10\pm2\%$), standing ($13\pm3\%$), lying ($13\pm8\%$), sleeping ($33\pm0\%$), other ($5\pm1\%$). From the complete 24 h data set for each hen, 10 different subsamples were taken: 1 and 2 min every 15 min, 30 min and h; 15 min every h and 2 h; 1 h and 2 h four times during the day. All samples were normalized to a daily behavioral profile, and each subsample was compared to the complete 24 h using a t-test. For time spent sleeping, the subsamples underestimated for sampling 1 min every h ($31\pm1\%$, $P<0.0001$), 1 h ($25\pm0\%$, $P<0.0001$) and 2 h ($26\pm1\%$, $P<0.0001$) four times through the day. There was no significant difference between subsamples and complete 24 h for any other observed behavior. These results indicate that any of the sampling schemes may be implemented (depending on sleeping behavior importance). Choosing the most efficient method for sampling in this scenario reduces the analysis time by 98%. Additionally, the electronic sensors were 98% accurate for detecting hen movement through the tunnels, increasing the meaningful data collected with minimal additional analysis time.

## Nesting behaviour of laying hens housed in enriched environment

*Pereira, Danilo Florentino[1], Batista, Edna dos Santos[1], Nagai, Douglas Ken[1], Costa, Michelly Aragão Guimarães[1] and Moura, Daniella Jorge de[2], [1]Univ Estadual Paulista – UNESP, Campus Experimental de Tupã, Av. Domingos da Costa Lopes, 780, 17602-496 – Tupã, SP, Brazil, [2]Univ Estadual de Campinas – UNICAMP, Faculdade de Engenharia Agrícola, Cidade Universitaria Zeferino Vaz, s/n, 13083-875 – Campinas, SP, Brazil; danilo@tupa.unesp.br*

Alternative systems for housing laying hens have been studied, and the furnished cages have shown the best results in the provision of welfare. However, there is no agreement on which the group size and stocking density are more suitable for the rearing conditions in Brazil. The aim of this study was to compare the behavior of laying hens in the two group sizes and two stocking density, wood shavings forage in an enriched environment. The experiment lasted four weeks, including the entire month of May 2010, and the first week was devoted to the adaptation of birds and the last 21 days include a production cycle of 21 days. We used 36 hens of Isa Brown strain, with 30 weeks of age at the start of the experiment. The birds were housed in two different sizes of groups (T1: 6 and T2: 12 birds) and in two densities: 774 cm$^2$/bird and 1440 cm$^2$/bird. In each treatment was placed a one nest built on wood with dimensions 40×40 cm. The behaviors of frequency of use and time spent in the nest were monitored by video cameras placed on the roof of the aviaries. Statistical analysis showed that there was statistical difference ($P < 0.05$) in frequency of use as the time spent in the nest, between treatments. It was found that the birds housed in group sizes of six birds and density of 774 cm$^2$/bird entered the nest significantly higher in both periods when compared with other treatments. In the afternoon it was observed that the length of stay in the nest in groups of six birds was significantly lower for groups of 12 birds. The study showed that the nesting behavior is very important for hens and stocking density and group size mainly affect such behavior. However, other studies should be conducted to confirm which densities and group sizes are most suitable.

**Effect of cage design on consistency of orientation and location during oviposition of laying hens**

*Engel, Joanna[1], Bont, Yoni[2] and Hemsworth, Paul[1], [1]University of Melbourne, School of Land and Environment, Parkville, VIC 3010, Australia, [2]Wageningen University, Animal Sciences, Building no. 531, Marijkeweg 40, 6709 PG Wageningen, Netherlands; jengel@student.unimelb.edu.au*

Laying hens often show a preference for a particular site at oviposition, in terms of on the floor or in a nest box. However, few studies have examined the location within the general site, e.g. location on the floor or in the nest box, or the orientation of the hen during oviposition. Within one replicate of a large study examining the effects on hen behaviour and physiology of floor space allowance (540 vs. 1650 $cm^2$/hen) and nest box access (present or absent) in a factorial design, observations were conducted from video records on 48 hens (12 from each of the 4 cage designs) on 14 days from 26 to 33 weeks of age, with an average of 13 observed eggs laid per hen. Floor areas of small cages were divided into 3 equal portions and large cages were divided into 9 equal portions with an additional 4th or 10th area if the cage contained a nest box, respectively. Each of these areas represented 180 $cm^2$/hen. Individual bird consistency of nesting site was based on a hen laying 80% of her eggs in the same location within the cage and consistency of orientation was based on a hen laying 80% of her eggs with the same orientation within the cage. Hens with access to a nest box were more consistent in their site of oviposition ($F_{1,4}$=54.715, P=0.002), as well as more consistent in their orientation during oviposition ($F_{1,4}$=47.220, P=0.002). Hens housed in smaller cages were more consistent in site of oviposition than hens in large cages ($F_{1,4}$=8.421, P=0.044). There were no significant interactions between floor space allowance and nest box access. Higher consistency in location and orientation during oviposition of hens in a nest box may be explained by a higher motivation to lay their egg in a nest box, but the nest may provide more protection from interference from other birds. Increased consistency in location of a hen during oviposition in a smaller cage may be due to space restrictions limiting interference. The role of housing design, and in turn stress, on egg laying behaviour requires further research. There are little data available on the effects of stress on consistency of egg laying behaviour.

## The effect of ramp slope on heart rate, handling and behaviour of market pigs at unloading

*Goumon, Sébastien[1,2], Bergeron, Renée[2,3] and Gonyou, Harold W.[1], [1]Prairie Swine Centre, Saskatoon, SK, Canada, [2]Laval University, Québec, QC, Canada, [3]University of Guelph, Guelph, ON, Canada; sebastien.goumon.1@ulaval.ca*

Loading and unloading have been identified as the most stressful events during transportation. Trailer design may play a significant role in the ease of handling in pigs. Our study aimed to assess the effect of ramp slope (0, 16.5, 21 or 25.5°) on the ease of handling, behaviour and heart rate of market pigs. Two hundred pigs were unloaded onto an apparatus designed to simulate unloading from the belly compartment of a pot belly trailer. The same handler was used to move pigs (groups of 10) with paddle and board. Heart rate (pigs and handler), unloading time, handler's interventions and behaviours of pigs were monitored. This study was designed as a randomized complete block design. Behaviour and handling datawere analyzed with a Kruskal Wallis test. Time and heart rate data were analyzed using a mixed model. The use of a 25.5° ramp led to a higher heart rate in the handler (25.5°:146.8, 21°:128.4, 16.5°: 129.4, 0°: 118.8 BPM, SE: 4.1), and a handling score indicating a more difficult handling ($P<0.01$). Unloading took longer (25.5°: 42.4, 21°: 18.1, 16.5°: 24.4, 0°: 14.3 s, SE: 5.9) and balking and vocalizations were more frequent with the 25.5° ramp ($P<0.01$), but the results were not significantly different from the 16.5° ramp ($P>0.05$). No significant effects of treatments on pig heart rate were found. Our results demonstrate that a 25.5° ramp makes handling more difficult and increases handler's heart rate.

## The comparison of two rating systems for Qualitative Behaviour Assessment in two situations

*Gutmann, Anke[1], Muellner, Beate[1], Leeb, Christine[1], Wemelsfelder, Francoise[2] and Winckler, Christoph[1], [1]University of Natural Resources and Life Sciences, Division of Livestock Sciences, Gregor-Mendel-Strasse 33, 1180 Vienna, Austria, [2]Scottish Agricultural College, Sustainable Livestock Systems, Bush Estate, Penicuik EH26 0PH, United Kingdom; christoph.winckler@boku.ac.at*

This study compares two different rating systems for Qualitative Behaviour Assessment (QBA) applied to two different situations: Free-Choice-Profiling (FCP), in which observers generated their own descriptors for scoring, and a Fixed-Terms-List (FTL), comprising 20 pre-defined descriptors based on the Welfare Quality® protocol. Video footage consisted of 20 brief clips showing loose-housed dairy cow herds in different contexts (herd-clips), and 20 clips showing different individual cows receiving social licking (SL-clips). 12 experienced observers assessed these clips, first using FCP to ensure bias-free generation of terms, and again one week later using FTL. The resulting 4 datasets were analysed using Generalised Procrustes Analysis, a multivariate technique that finds consensus dimensions and attributes scores to each video clip on these dimensions. Observer agreement was significant in all 4 analyses ($P<0.001$). For the herd-clips, FCP dimension 1 (FCP1) ranged from restless/tense to calm/relaxed and explained 39.6% of the variation between clips, while FTL dimension 1 (FTL1) ranged from agitated/stressed to calm/relaxed, explaining 42.0%. FCP2 ranged from indecisive/expectant to active/confident explaining 10.1% of the variation, while FTL2 ranged from bored/apathetic to lively/playful explaining 15.2%. The Pearson correlation between FCP and FTL clip scores was 0.96 ($P<0.001$) for dimension 1, and 0.67 ($P<0.001$) for dimension 2. For the SL-clips, FCP1 ranged from inviting/appreciative to indifferent/passive and explained 54.2% of the variation, while FTL1 ranged from inviting/appreciative to indifferent/bored, explaining 51.7%. FCP2 ranged from reluctant/pressured to calm/relaxed explaining 10.7% of the variation, while FTL2 ranged from agitated/uneasy to content/calm explaining 13.2%. The Pearson correlation between FCP and FTL clip scores was 0.96 ($P<0.001$) for dimension 1, and 0.87 ($P<0.001$) for dimension 2. These results demonstrate a high consistency between different QBA rating systems and provide further support for the reliability of QBA as a 'whole-animal' assessment technique.

## A comparison of three cattle temperament assessment methods

Sant'anna, Aline [1,2], Paranhos Da Costa, Mateus[1], Rueda, Paola[1,3], Soares, Désirée[1,3] and Wemelsfelder, Francoise [4], [1]Faculdade de Ciências Agrárias e Veterinárias, UNESP, Departamento de Zootecnia, Jaboticabal, 14884-900, Brazil, [2]Pós-Graduação em Genética e Melhoramento Animal, Jaboticabal, 14884-900, Brazil, [3]Pós-Graduação em Zootecnia, Jaboticabal, 14884-900, Brazil, [4]Scottish Agricultural College, Edinburgh, EH9 3JG, United Kingdom; ac_santanna@yahoo.com.br

The aim of this study was to compare three methods to assess cattle temperament. Data from 335 Nellore young bulls were recorded, measuring: (1) score of movement in the crush (MOV), from 1 (no movement) to 4 (movements frequent and vigorous); (2) flight speed (FS), recording the speed that an animal exit a crush; and (3) flight distance (FD), using scores from 1 (when an animal allows to be touched) to 5 (when an animal shows aggression towards the observer; this measure was done with the animal kept in a 30 $m^2$ pen). The qualitative behaviour assessment method (QBA) was used as a reference to explain the variation of each method's approach to cattle temperament. It was adapted to assess cattle temperament using 12 terms (active, relaxed, fearful, agitated, calm, attentive, positively occupied, curious, irritable, apathetic, happy and stressed). The observer indicates his qualitative assessment of an animal's expression by scoring each term on a line of 125 mm, where the minimum represents absence of the term expression, and maximum an intense manifestation of it. Pearson's coefficients of correlation were estimated to assess the association between the tests and between each test and each QBA term, assuming P<0.01. Significant correlations were found between MOV and FS (0.194) and FS and FD (0.194), but not between MOV and FD (0.008). Regarding the QBA terms: MOV was significantly correlated with the terms active (0.240), calm (-0.216) and relaxed (-0.200); while FS was significantly correlated with active (0.555), agitated (0.501) and attentive (0.484); and FD with happy (-0.288), calm (-0.236) and apathetic (-0.222). Most of the correlation coefficients were low, only FS showed moderate values with some QBA terms. We conclude that there are variations in the methods' approach to cattle temperament; with FS addressing more the expressions of activity and agitation. Probably the differences in aspects of temperament approached are due to the context in which each test is applied. Financial Support: FAPESP.

## When a duck leads, do others follow?

*Liste, Guiomar[1], Asher, Lucy[2], Kirkden, Richard D[1] and Broom, Donald M[1], [1]University of Cambridge, Department of Veterinary Medicine, Madingley Road, CB3 0ES Cambridge, United Kingdom, [2]University of Nottingham, School of Veterinary Medicine and Science, Sutton Bonington Campus, LE12 5RD Loughborough, United Kingdom; gl318@cam.ac.uk*

This study forms part of a wider project investigating the provision of open water sources for farmed ducks. Bathing behaviour is thought to have a strong social component which could have relevant consequences when offered in a commercial situation. A test was designed to determine preference between open water sources of different depths. Although no significant preference could be identified, the data were also used to investigate group behaviour in ducks, especially during bathing. 16 groups of Cherry Valley Pekin ducks were housed in pens with concrete floors and straw bedding from 21 days of age. Pools and drinkers were located on a raised slatted floor area along one side of the pen, accessed by a concrete ramp. Each group consisted of 4 ducks with access to 4 resource areas: POOLS ($2.20 m^2$), AROUND POOLS ($3.60 m^2$), DRINKERS ($3.00 m^2$) and STRAW ($18.00 m^2$). Ducks were individually identified and 24 h video recordings were made using CCTV cameras at 29, 34, 36, 41, 43 and 48 days. Behaviour was continuously observed and transitions between areas were scored using Observer XT9. A duck was considered to initiate a movement if it was the $1^{st}$ to move to an unoccupied area. The 4 ducks in each pen were categorised according to the % of times they initiated movements (a duck was considered leader if it initiated movements on significantly more than 25% of occasions). In 72% of the groups one duck consistently led for all videos watched. This individual initiated $40(\pm0.1)$% of all movements made. Percentages were arcsine transformed and GLMM was performed with 'area' and 'leader status' as fixed factors. This confirmed that the consistent leader initiated a significantly higher proportion of movements than any other duck ($P<0.001$) and that this difference in leadership was more pronounced ($P<0.03$) when movements were made into POOLS. Hence, ducks behaved socially by following a consistent leader, especially when moving towards an open water source. This should be considered when assessing group vs. individual preferences in the future. The existence of leaders and the manner in which this affects the use of pools in a commercial situation (monopoly, defence) needs further investigation.

### Different behavioural reactions to open field in growing broiler and layer chickens

*Baranyiová, Eva[1] and Balážová, Linda[2], [1]Czech University of Life Sciences, Institute for Tropics and Subtropics, Kamýcká 129, 16521 Prague, Czech Republic, [2]Veterinary Clinic G and W, Vyšehradská 9, 851 06 Bratislava, Slovakia (Slovak Republic); ebaranyi@seznam.cz*

Selection for high meat or egg production lead also to changes in the behaviour, especially locomotor activity, of fowl. The open field test (OFT) has been used to assess fear responses in chickens. Our objective was to compare the behaviour of growing broilers (ROSS 308) and layers (ISA Brown) in a stress situation, namely, during repeated OFT. Three-day-old males (n=12) and females (n=12) of each strain were placed individually in an OF for 10-min periods. Six tests were carried out and video-recorded at weekly intervals until post-hatching day 42. Behaviour was analysed using ANOVA for repeated measures with Tukey post-hoc test. Layer and broiler chicks differed in time of occurrence of their reactions to repeated OFT: horizontal locomotor activity (HLA), visual orientation (turning the head, looking around, VO), vocalization and comfort behaviour. In addition, layers showed vertical locomotor activity (attempts to fly), pecking and freezing but without significant changes. Broiler males decreased the above behaviours (HLA: $P<0.05$; VO and vocalization: $P<0.001$) in the second OFT. Females decreased their HLA ($P<0.05$) and VO ($P<0.001$) in OFT 2 but vocalization time decreased later, in OFT 3 ($P<0.001$). Sex differences occurred in VO ($P<0.01$) and vocalization ($P<0.05$). Both sexes showed longer comfort behaviour as soon as in OFT 2 ($P<0.001$). Layer males and females decreased their HLA ($P<0.05$) later, in OFT 4, and vocalization in OFT 5 and 6 (males: $P<0.01$, $P<0.001$; females $P<0.05$). Their VO time increasedin OFT 6 (males: $P<0.01$; females: $P<0.01$). No changes in comfort behaviour occurred. In conclusion: Layers were more active in OFT than broilers: they moved (HLA) more (males: $P<0.001$; females: $P<0.05$, vocalized longer (males: $P<0.05$; females: $P<0.001$), and showed longer VO (males: $P<0.001$; females: $P<0.001$) but spent less time in comfort behaviour. Thus in a stress situation of OFT, the behaviour of both strains showed differences that occurred in early ontogeny. Supported by IGA VFUB grant 42/2007.

## The effects of using different levels and sources of zinc with wire vs. solid sided cages on laying hen feather quality

*Purdum, Sheila E., Aljamal, Alia A. and Krishnan, Pradeep, UNL, Animal Science, 3801 Fair St., 68583-0908 C206b, USA*

The objective of this study was to investigate the effects of using different sources and levels of zinc (Zn) as well as different caging systems (wire vs. solid sided) on egg quality parameters, feather score and welfare of white leghorn hens. Two hundred forty Hy-Line W-36 laying hens were fed the dietary treatments for 30 weeks from 45 to 75 wks of age. Hens were fed a corn-soybean meal basal diet with four different combinations of Zn sources and levels: (treatment 1) 40 ppm organic Zn (Availa Zinc), (treatment 2) 60 ppm inorganic Zn (Zn sulfate), (treatment 3) 40 ppm Availa Zinc+ 20 ppm inorganic Zn, and (treatment 4) 40 inorganic Zn+ 40 Availa Zn. Hens were assigned to a total of 48 cages with 12 replicates/dietary treatment with two types of cages: Farmer Automatic Cages (solid sided cages) with 4 hens/cage and 500 sq cm/hen, Big Dutchman cages (wire sided cages) with 6 hens/cage and 600 sq cm/hen. Cages were assigned in a randomized complete block design. Feed intake and egg production were measured daily. Body weights were taken monthly. Feather scoring was done on a weekly basis. A 5-point scale was used to score feathers of neck, breast, wing, back, tail, and cloaca of hens, where 1 = fully feathered, 2 = ruffled, no naked spots; 3 = naked spots up to 5 cm at the widest part; 4 = naked spots greater than 5 cm wide; and 5 = very poorly feathered with naked spots and injury to the skin (Webster and Hurnik, Poult. Sci. 69: 2118-2121, 1990). Dietary treatments had no significant effect on egg production variables. There was an increase (P=0.0102) in body weight of hens housed in solid sided cages (1.57 kg) compared to those housed in wire sided cages (1.46 kg). Eggs produced in solid sided cages showed significantly higher yolk solids % compared to the wire sided cages (P=0.0427). There was no difference between dietary treatments on feather score. However a significant difference was observed in feather score for breast, wing, back, tail and cloaca with the cage type; the feather scores were consistently better for hens housed in the solid type cages compared to the wire sided ones (P<0.03). The results of the study showed that solid sided cages significantly improved welfare of the hen by weight gain, egg quality, and feather quality.

## Stress following caging of shelter cats (*Felis silvestrus catus*)

*Ellis, Jacklyn J, Protopapadaki, Vasiliki, Stryhn, Henrik, Spears, Jonathan and Cockram, Michael S, Atlantic Veterinary College, University of Prince Edward Island, Sir James Dunn Animal Welfare Centre, Department of Health Management, 550 University Ave, Charlottetown, PE, C1A 4P3, Canada; jjellis@upei.ca*

This study investigated methods of identifying stress in caged cats and the changes in physiological and behavioural signs of stress coinciding with habituation to the environment, and provides a behavioural activity budget of cats during caging. Six cats from a local shelter were kept in cages for 4 weeks and videotaped 24 h/day under 12 h light and 12 h dark (plus infrared lights). Continuous focal observations of the activity, location in the cage, and posture were recorded for one 24 h period/week/cat. Qualitative Cat-Stress-Scores (CSS) were recorded daily. All faecal samples produced during the study period were collected for analysis of faecal glucocorticoid metabolites (FGM). Effects of time were examined using GLMs and Bonferroni pairwise comparisons. Where data could not be transformed to achieve normality, Friedman's and one sample sign tests were used. Loge FGM data showed a declining trend across days in 4 of 6 cats (P?0.1), but there was large between-cat variation. CSS declined significantly across weeks (Friedman's test: df=4, P<0.01), with a significant decrease (P<0.01) from week 1 (median=2.4, IQR=0.91, n=6) to week 2, (median=2.1, IQR=0.36, n=6). The daily percentages of time spent eating and grooming were significantly different over time (both P<0.01). Eating increased significantly (P<0.01) from week 1 (mean=1.4, SD=0.91, n=6) to week 2 (mean=2.8, SD=0.55, n=6), while grooming decreased significantly (P<0.01) from week 1 (mean=5.2, SD=2.20, n=6) to week 2 (mean=2.3, SD=0.86, n=6). After week 1, cats spent most of their day inactive (mean=88%, SD=2.3, n=24), with on the shelf as their primary location (median=45%, IQR=73.2, n=24), and with lying head down as their primary posture (mean=55%, SD=10.8, n=24). These behavioural categories were measured independently of each other. As cats were located on the shelf almost half of the time this suggests that it may be a resource of value to the cat, and that its inclusion in enclosure design could be important. Physiological, as well as quantitative and qualitative behavioural data suggested an initial stress response to caging that was followed by a gradual decline with time.

**Bathing behavior of captive Orange-Winged Amazon parrots (*Amazona amazonica*)**

*Murphy, Shannon, Braun, Jerome and Millam, James, University of California, Davis, One Shields Avenue, Davis, CA 95616, USA; smmurph@ucdavis.edu*

Orange-winged Amazon parrots (OWA) rain-bathe, a style of water-bathing typified by a series of stereotyped bathing postures. In captivity, rain can be simulated and rain-bathing (RB) induced by spraying with a hand-held mister. Our purpose was to: (1) describe the postures and sequence of postures of OWA during RB; (2) determine the impact of RB on activity budgets of resting, feeding and preening; and (3) characterize the temporal patterning of RB across the day and test whether refractoriness occurs to spray-induction of RB. In Exp. 1, RB was induced in 12 OWA and RB postures and sequences of postures video recorded. Eight distinct RB postures were identified, including a drying posture displayed after rain ceases. Exp. 2 used continuous sampling to observe activity budgets of OWA (N=12) during three, 2-hr observation periods (morning, midday and afternoon). Analysis by mixed effects models found that OWA spend ~90% of their time resting, feeding and preening. A crepuscular pattern of activity was evident (P=0.01), with more feeding observed in the morning and afternoon and more resting in the midday; preening was stable across the day occupying ~19% of time. In Exp. 3 spray-induced RB was observed at midday and activity budgets recorded 1 hr before and 1 hr after a 15-min spray period. Resting, feeding and preening comprised >85% of time in both observation periods; only preening decreased between the observation periods (P=0.0045), and no other changes were significant. OWA spent ~9 min in RB and had a ~2 min latency-to-bathe during the 15-min spray period. In Exp. 4, OWA (N=11) were sprayed for 20 min on Day 0, then split into three groups and sprayed after 2-, 4- or 6-day refractory intervals (RI). RB time was increased (P=0.008) and latency-to-bathe decreased (P =0.003) by RI-6 d. Lastly (Exp. 5), OWA (N=11) sprayed for 20 min in the morning, midday or afternoon bathed for ~11 min in the morning and less thereafter (P=0.01), while latency-to-bathe was ~ 1.5 min and unchanged across the day. These results are consistent with the very limited literature describing RB in other wild parrot species and suggest that spraying may be incorporated as a routine environmental enrichment to induce RB in captive Amazon parrots.

**Decline in aggression in cotton rats through the use of enrichment**

*Neubauer, Teresa, Zabriskie, Ryan and Buckmaster, Cindy, Baylor College of Medicine, Center for Comparative Medicine, One Baylor Plaza, Houston, Texas 77030, USA; neubauer@bcm.edu*

The cotton rat (*Sigmodon hispidus*) presents a unique susceptibility toward human pathogens and is currently used in studies of human respiratory syncytial virus, adenovirus and parainfluenza virus. Though very valuable, this animal model has earned a bad reputation with husbandry, veterinary and research personnel. Cotton rats tend to be hyperactive and aggressive in the lab setting and are likely to leap from their cages or bite when handled. Frequent fighting between cage mates, particularly males, is often brutal and deadly, compromising study results. Our researchers were very concerned about the aggression and fighting in their colony and asked us for a solution. An internet search revealed nothing helpful, so we suggested the addition of cardboard tubes (3.2" × 5" × 0.30") to the cages for shelter and enrichment. 156 cotton rats were pair housed. Paired males received two tubes to prevent competition while paired female and breeder cages received one tube. Cotton rats received morning health observations 365 days a year. All incidences of fight wounds were recorded, regardless of severity, and recorded in a database in order to track animals over time. In one year, fight wounds decreased by 77%. A comparison of fight wound incidents with and without the enrichment specified revealed a highly significant reduction in injury when the enrichment was in place (P-value = 0.0031). A qualitative change in the animals' behavior was also noted by our technicians, veterinarians and researchers. Prior to the addition of the tubes, it was necessary to handle the rats wearing heavy leather gloves or HexArmor gloves for protection from biting. We now wear latex gloves during cage changing with no more fear of biting than from other common rodents.

## Effects of stressors on the behavior and physiology of domestic cats

*Stella, Judi, Croney, Candace and Buffington, Tony, The Ohio State University, College of Veterinary Medicine, 1920 Coffey Rd, Columbus, OH 43210, USA; stella.7@osu.edu*

Feline interstitial cystitis (FIC) is a painful bladder disease. Cats with FIC have chronic or recurrent lower urinary signs (LUTS) and other comorbid disorders that are exacerbated by stressors. The aim of this study was to evaluate behavioral and physiological responses of healthy (H) cats and cats diagnosed with FIC after a three day stressor was imposed. Ten healthy cats and 18 cats with FIC were housed at the OSU Veterinary Medical Center vivarium. All cats were singly housed in enriched cages for at least one year prior to the experiment. Cats had daily play time outside of the cage, treats, auditory enrichment, and socialization time with one author (JS). The daily husbandry schedule was maintained at a consistent time of day and cats were cared for by two familiar caretakers. During test days, cats were exposed to multiple simultaneous stressors considered to be minor disruptions that would occur under normal vivarium conditions. Stressors included multiple caretakers unfamiliar to the cats, an inconsistent husbandry schedule, and discontinuation of play time, treats, auditory enrichment, and positive human-animal interactions. Cage conditions and food were kept consistent throughout the study. On the morning before and after the stressor, blood samples were collected for measurement of cortisol, leukocytes, lymphocytes, neutrophils, and the cytokines IL-1, IL-6, and TNF-a. Sickness behaviors (SB) including vomiting, diarrhea, anorexia or decreased food and water intake, fever, lethargy, somnolence, enhanced pain-like behaviors, decreased general activity, body care activities (grooming), and social interactions were recorded daily. Both healthy cats and cats with FIC had statistically significant (paired t-test) increases from baseline measures in SB during the stress period (H cats P=0.002; FIC P=0.0001). The stress period also resulted in a significant decrease in lymphocytes in FIC, but not healthy cats (H cats P=0.98; FIC P=0.009). No differences were observed for any other parameters. Overall, the short term stressors led to a significant increase in SB in both healthy cats and cats with FIC, whereas lymphopenia occurred only in FIC cats. Daily monitoring of cats for SB may be a noninvasive and reliable way to assess stress responses and overall welfare of cats.

## Dairy cattle preferences for feed bunks with or without sprinklers in summer

*Chen, Jennifer M.[1], Schutz, Karin E.[2] and Tucker, Cassandra B.[1], [1]UC Davis, Davis, CA, USA, [2]AgResearch, Ltd, Hamilton, New Zealand; jmchen@ucdavis.edu*

Sprinklers effectively reduce heat load in dairy cattle, but elicit variable behavioral responses: in some studies, cattle readily use sprinklers, while in other studies cattle either avoid or show no preference for them. In the United States, a common way to cool cows with water is with nozzles mounted over the feed bunk that intermittently spray (i.e. 5 min on, 10 min off, as in this experiment) animals' backs while they eat. The objective of this study was to assess the use (phase 1) and preferences (phase 2) for this type of sprinkler system. Lactating cows were tested in groups of 3 animals (n=8 groups) in the summer (mean 24 h temperature 28.5±3.0 °C). Behavioral data were collected with 5-min scan sampling and a paired or a 1-sample t-test was used for all statistical comparisons. In phase 1 of the study, cows were fed from shaded bunks with or without sprinklers for 2 d each, with the order of exposure balanced in a cross-over design. Cows spent more time at the feed bunk fitted with sprinklers both feeding (sprinkler vs. non-sprinkler: 207 vs. 149 min/24 h, SE: 7 min/24 h, P<0.01) and standing without feeding (259 vs. 135 min/24 h, SE: 19 min/24 h, P<0.01). When using the sprinklers, cows protected their heads from being sprayed directly: when they were not feeding, cows were more likely to put their heads past the feed bunk barrier when the sprinklers were on (on vs. off: 78.3 vs. 59.4% of time standing without feeding, SE: 2.8%, P<0.01). Average body temperature was lower in the sprinkler treatment (38.8 vs. 39.2 °C, SE: 0.1 °C, P=0.01). In phase 2, *ad libitum* access to feed was provided in both treatments for 5 d and cow preference was assessed. All groups preferred the feed bunk with sprinklers while both feeding (69.4 vs. 30.6% of time, SE: 15.4%, P<0.01) and standing without feeding (84.5 vs. 15.5% of time, SE: 7.7%, P<0.01); these preferences were seen regardless of whether the water was on or off. In conclusion, cows spent more time at feed bunks with sprinklers compared to those without this resource, and when given access to both, preferred to spend time at feed bunks with sprinklers. To better understand the differences in behavioral responses to sprinklers between studies, other factors such as previous experience and physical aspects of the sprinkler system (e.g. droplet size) need to be explored.

## Effects of alternative housing and feeding systems on the behavior and performance of dairy heifer calves

*Pempek, J.A., Eastridge, M.L., Botheras, N.A., Croney, C.C. and Bowen, W.S., The Ohio State University, Animal Sciences, 2029 Fyffe Rd., Columbus, OH 43210, USA; jessica.pempek@gmail.com*

Most calves in the dairy industry are housed individually prior to weaning. However, this type of housing limits the calves' ability to display social behavior, which may impede development of normal social responses. Individual housing is often preferred to minimize undesirable behaviors such as cross-sucking. Previous studies have indicated that if calves are fed with a bottle instead of a bucket, these undesirable behaviors may be reduced. In this experiment, forty-eight female Holstein calves were assigned to the following treatments at $6\pm3$ d of age and monitored for approximately 9 wk: individual housing fed with a bucket, individual housing fed with a bottle, paired housing fed with a bucket, or paired housing fed with a bottle. Calves were housed in pens separated by wire panels, through which all calves could cross-suck on neighboring calves. Milk was fed via bucket or bottle twice/d (6 l/d). Calves had *ad libitum* access to calf-starter and water. Gradual weaning commenced at wk 6 by reducing the calves' milk allowance by 2 l/wk. Calves were weaned at the beginning of wk 8. Grain consumption and body weight were monitored on a weekly basis and wither height measured at the beginning and end of the experiment. Behavior was video recorded for 2 h both in the morning and afternoon for 16 calves during wk 1 and scored by scan sampling 1 frame every 10 min. There was no effect of housing or feeding method on posture, self-grooming, or play behavior. During the 20 min following milk delivery watched continuously, calves housed individually spent more time engaged in non-nutritive oral behavior (Individual: $19.3\pm2.9\%$; Paired: $7.0\pm2.9\%$; P=0.02) and less time cross-sucking ($12.4\pm5.5\%$; $31.0\pm5.5\%$; P=0.04) than calves housed in pairs. Bottle-fed calves spent more time feeding than those fed with a bucket (Bottle: $39.5\pm1.0\%$; Bucket: $25.9\pm1.0\%$; P<0.0001). There were no significant differences for housing or feeding method on total dry matter intake or change in body weight or wither height. These results indicate that housing heifer calves in pairs allows for social interactions without being detrimental to their performance. However, feeding with a bottle did not reduce cross-sucking.

**Evaluation of the relationship between temperament, time spent at the feed trough and weight gain of finishing weight feedlot cattle**

*Soares, Désirée Ribeiro[1], Schwartzkopf-Genswein, Karen[2], Sant'anna, Aline Cristina[1], Valente, Tiago da Silva[1], Rueda, Paola Moretti[1], Cyrillo, Joslaine Noely dos Santos Gonçalves[3] and Paranhos Da Costa, Mateus José Rodrigues[1], [1]Programa de Pós Graduação FCAV/UNESP – Grupo ETCO, Zootecnia, Prof. Paulo Donatto Castellane, Jaboticabal,SP, 14884-900, Brazil, [2]Agriculture and Agri-Food Canada, Lethbridge Research Centre, 5403 1 Avenue South, Lethbridge, Alberta, P.O. Box 3000, Canada, [3]Instituto de Zootecnia, Estação Experimental, Rodovia Carlos Tonanni Km 94, Sertãozinho-SP, 14160-900, Brazil; soares.desiree@gmail.com*

The aim of this study was to determine the relationship between beef cattle temperament, time spent at the feed trough (TST) and average daily gain (ADG) in feedlots. TST was monitored for 10 d (07:00 h to 18:00 h) using instantaneous scan sampling conducted at 5 min intervals. Fifty-three bulls (35 Nelore and 18 Nelore cross with an average age of 30 + 3 mo.) were observed within one feedlot pen (5. 246 m$^2$); the feed bunk (30 m long) was placed in the middle of the pen, allowing the animals to feed at same time and from both sides. The animals were fed a diet (fed basis) consisting of 12.3 kg of sugar cane silage and 8.2 kg concentrate/animal/day (31.2% ground corn, 6.7% soybean meal, 0.5% urea, 1.6% mineral core – Nutron 555AJ) delivered at 09:00 h and 16:00 h daily. Temperament was assessed on d 1, d 27 and d 54 at the end of the fattening period by measuring flight distance (FD: proximity (m) to which a stock person could come to an individual animal before it would move away) and flight speed (FS: speed (m/s) at which the animal exited a handling chute). The ADG was calculated using animal weights obtained on d 1 and d 54 at the end of the fattening period. Pearson's correlation coefficients were estimated for all variables. TST was significantly correlated to FD (r=-0.51) and FS (r=-0.37). However, there was no correlation (P>0.05) between ADG and indicators of temperament (FS: r=-0.03 / FD: r=0.09), nor with TST (r=0.10). Based on these results we conclude that in spite of less reactive animals spending more time at trough, there is no association with weight gain. This may be explained by the fact that greater time at the feed trough does not necessarily mean the animals are consuming more feed. Future studies should focus on understanding the relationship between temperament and feed intake.

**Effects of providing a roof and locating food in an outside yard on behaviour of sheep kept in winter conditions**

Jørgensen, Grete Helen Meisfjord[1] and Bøe, Knut Egil [2], [1]Bioforsk Norwegian Institute for Agricultural and Environmental Research, Tjøtta, 8860 Tjøtta, Norway, [2]Norwegian University of Life Sciences, Department of Animal and Aquacultural Sciences, P.O. Box 5003, 1432 Ås, Norway; grete.jorgensen@bioforsk.no

We aimed to investigate the effect of roof cover and location of feed on sheep's use of an outdoor yard under different weather conditions. A 2 x 2 factorial experiment was conducted with roof covering of outdoor yard (yes or no) and location of feed (indoors or outdoors) in four different pens, each with one of four possible combinations of these factors. The outdoor temperature varied from +16.2 to -27.9 °C in and the precipitation per day was from 0 to 38 mm in the experimental period. Twenty adult ewes of the Norwegian White breed were randomly allotted to 4 groups with 5 animals. The ewes were fully fleeced and groups were rotated between treatment pens every week from November 2009 to March 2010. Twenty-four hour video recordings were performed once a week and general behaviours (standing, lying, feeding) and location (outdoors or indoors) was scored for each individual using instantaneous sampling every 15 minutes. A mixed model of analysis of variance was applied with group specified as a random effect. The interaction between roof and location of feed was also examined. The weather did not seem to have any large effect on general behaviours. Sheep spent more time in outdoor yards that were covered with a roof (43.8±1.3 vs. 36.3±1.4% of tot. obs.; $F_{1,10}$=7.1; P<0.05), and they also rested more outdoors in such yards (24.2±1.4%) compared to in yards that did not have a roof cover (15.5±1.2%; $F_{1,10}$=12.9; P<0.01). Locating the feed outdoors increased the time spent resting indoors (47.4±1.5% vs. 31.6±1.5%; $F_{1,10}$=46.2; P<0.0001), indicating that if a dry and comfortable resting area is offered indoors, the feed should be located in the outdoor yard. Utilizing an outdoor yard as part of the total area may be a cost effective way of meeting new space regulations for sheep production.

## Evaluation of conventional and large group auto-sort systems for grow/finish pigs

*Brown, Jennifer, Hayne, Stephanie, Samarakone, Thusith, Street, Brandy and Gonyou, Harold,
Prairie Swine Centre, Box 21057, 2105 8th Street E., Saskatoon, S7H 5N9, Saskatchewan, Canada;
jennifer.brown@usask.ca*

Managing pigs in large groups allows the implementation of new technologies that would not be
feasible in small groups, however, some studies have shown an initial reduction in performance
following the formation of large groups. We hypothesized that feeding behaviour in large
groups is negatively affected by the design of food courts and compared feeding behaviour
and productivity in 2 large group pen designs and in a conventional housing system. Pigs were
housed in conventional small pens (C: 18 pigs/pen, n= 24), large groups with feeders in the
centre of the food court (LGC: 250 pigs/room, n= 4), and large groups with feeders on the
periphery of the food court (LGP: 250 pigs/room, n= 4), all on fully slatted floors. Pigs entered
the rooms at 10 weeks of age, with LGC and LGP pigs being introduced directly to the food
court area. Feeding behaviour was recorded at 5 min intervals using time-lapse cameras and
performance and injuries were recorded until pigs reached market weight. Effects of housing
treatment were assessed using mixed model ANOVA and Chi square analysis. No difference
in average daily gain was found among the treatments (P>0.05), although large-group auto-
sort pens emptied more slowly in the latter part of the marketing period (P<0.01). Pigs in C
had a higher incidence of tail-biting than did large groups (C: 18% vs. LGC and LGP: 1%;
P<0.01), but large group rooms had a higher incidence of non-thrifty pigs (pigs removed for
reasons excluding circo-virus; C: 3.9% vs. LGC: 9.7% and LGP: 7.6%; P<0.01). Comparison
of the 2 food court arrangements showed that pigs had similar numbers of meals (LGC: 5.0
±1.5, LGP: 4.7 ±1.8 meals/day; P>0.05), accessed similar numbers of feeders (LGC: 5.6 ±2.1,
LGP 5.5 ±1.9 feeders/day; P>0.05), and showed a typical crepuscular feeding pattern. Carcass
measures showed that pigs in LGC had more backfat than pigs from conventional pens (LGC:
18.2 ±0.26 mm vs. C: 16.9 ±0.26 mm; P<0.01), while LGP pigs were intermediate (17.5 ±0.38
mm). In conclusion, pigs in large groups did not show the initial reduction in performance
seen in previous studies, possibly due to the fact that pigs were introduced directly to the food
court. Tail biting was significantly reduced in large groups suggesting that, if properly managed,
large groups can benefit welfare.

**Effects of ractopamine on stress-related hormone levels of purebred Berkshire swine**

*Betts, Katherine S., Moeller, Steven J., Zerby, Henry N., Crawford, Sara M., Cressman, Michael D. and Bishop, Megan J., The Ohio State University, Animal Sciences, 2029 Fyffe Court, Columbus, Ohio 43210, USA; betts.75@buckeyemail.osu.edu*

The objective of this study was to evaluate the effects of a 28 d pre-harvest ractopamine (RAC) feeding program on cortisol levels in purebred Berkshire pigs (n=62) using a randomized complete block design with three treatments (Control (C) 0 ppm; RAC5, 5.0 ppm; RAC10; 10 ppm) in duplicate. Littermates were randomly assigned to each RAC treatment. Pigs were weighed on test at an average of 92 kg for all 3 treatments. Pigs were housed in the same room in adjacent pens fitted with partially slatted floors and provided *ad libitum* access to feed and water. Salivary cortisol samples were collected from each pig at 0, 7, 14, 21, and 28 d of the feeding. Saliva samples were collected in a standard handling aisle, allowing a maximum of 120 s in the holding area. One individual, familiar to the pigs, collected the samples, with collection time recorded and a difficulty score assessed (1 to 10 scale, i.e. 1=sample easily collected and 10=no sample was obtained). Pigs were weighed off test (C=119 kg; RAC5=122 kg; RAC10=123 kg), transported to the harvest facility, rested 15 h, and harvested using an electrical stun followed by exsanguination. At 24 h postmortem, carcasses were ribbed between the $10^{th}$ and $11^{th}$ rib and fresh loin visual color (VC), marbling (M), firmness (F), wetness (W) scores, pH and instrumental Minolta L* were recorded. Analyses included a fixed effect of treatment and a random effect of litter within replicate. Baseline (0 d) cortisol levels did not differ among treatment groups. Salivary cortisol on d7 was greater for RAC10 (2.5571 ng/ml) when compared with C (1.9184 ng/ml; P=0.0767) and RAC5 (1.5942 ng/ml; P<0.01). Cortisol levels were not different among treatments at 14, 21 and 28 d. Correlations between weekly salivary cortisol measures, within and across RAC treatments, were not different from zero, indicating a low level of repeatability. Day 28 salivary cortisol level was not correlated with any measure of pork quality, indicating that the addition of RAC at 5.0 or 10.0 ppm did not negatively influence fresh pork quality when compared with pigs fed a control diet following a 28 d feeding period. Ractopamine improved efficiency and growth rate while maintaining muscle quality with no impact on salivary cortisol levels.

## Use of flavour association and preference tests in pigs to assess the palatability of pea diets

*Rajendram, Janardhanan, Beaulieu, Denise and Gonyou, Harold W., Prairie Swine Centre, P.O. Box 21057, 2105 8th Street, Saskatoon, S7H 5N9, Canada; raj152@mail.usask.ca*

The high net energy and digestible lysine content of field peas should allow for their incorporation into a wide range of diets. However, primarily because of concerns over palatability, usage for swine is limited. The objectives of this study were (1) to compare the palatability of diets varying in pea content, and (2), to assess whether pigs' aversion to peas is due to a taste effect or a post-ingestive effect. Experiment 1 examined the effect of level of pea inclusion on feed consumption. Fifty mixed gender pigs (9 weeks old) were fed 5 treatment diets (basal soy diet, 20, 40, 60% peas, canola control) in a completely randomized design for 10 days. The peas were added at the expense of wheat and soy to the soy diet. The canola diet was required to evaluate the response to a novel diet. The pigs had access to the diets for a 4-hour period each day. The data were analysed using mixed model ANOVA in SAS. Consumption levels for either the first or final 3 days were not different for either 20, 40 or 60% pea diets, compared to the soy basal or canola control diets (P>0.10). Experiment 2 was designed to examine post-ingestive feedback effects of peas. Twenty mixed gender pigs (8 weeks old) were fed either a 60% pea or a 10% canola diet on alternate days for 10 days. The diets were flavoured with 6 gm/kg of either orange or grape Kool-Aid$^{TM}$, with 10 pigs receiving peas/grape and canola/orange, and 10 receiving peas/orange and canola/grape on alternate days. Pigs were then presented with both an orange flavoured and grape flavoured basal diet to assess flavour preferences. The assumption is that if a diet produced negative post-ingestive feedback it would reduce feed consumption of the associated flavour during preference testing. Pigs did not exhibit a preference for either grape over orange flavour (P=0.46). This was irrespective of which diet had previously been associated with grape flavouring, as evidenced by the similarity in feed intake between the two diets (0.88±0.3 and 0.89±0.2 kg for pea and canola-based diets,respectively; mean ± SD, P=0.94). In conclusion, peas used in this study did not have any palatability issues suggesting that pea inclusion in diets does not affect feed intake. Further studies are being conducted to evaluate the effect of peas on the feeding behaviour of pigs.

## The influence of Acid-Buf™ mineral supplement on behaviour and salivary cortisol concentrations of uncastrated male and female growing pigs

*Boyle, Laura[1], O'gorman, Denise[2], Taylor, Stephen[2] and O'driscoll, Keelin[1], [1]Teagasc, Animal and Grassland Research and Innovation Centre, Moorepark, Fermoy, Co. Cork, Ireland, [2]Celtic Sea Minerals, Currabinny, Carrigaline, Co. Cork, Ireland; laura.boyle@teagasc.ie*

Magnesium can positively affect an animal's response and resistance to stress. The aim of this study was to evaluate whether supplementation with Acid-Buf™ (AB) which contains bioavailable Mg would reduce aggression, harmful and sexual behaviour and salivary cortisol concentrations in growing pigs. At weaning (28 d) 448 piglets were assigned to either Control (CL; Mg 0.16%) or AB (Mg 0.18%) diets in single sex groups of 14. Four wks later pigs (17.4±6.4 kg) were blocked according to weight and back-test scores. Seven pigs from each pen were mixed into fully slatted pens with 7 from another of the same sex and diet: CL male, AB male, CL female and AB female (n=4 of each). Pig behaviour was recorded 1 day/wk for 9 wks until pigs were c. 60 kg. All occurrences of the following behaviours were recorded in each pen during 8×2 min periods: fight, head-knock, bite (aggression); tail/ear in mouth, belly nosing (harmful); mount (sexual). The behaviour of 4 focal pigs/pen was recorded continuously for 2×5 min periods (am/pm) on the same day. Saliva was collected once/wk at 10:00 by allowing the 4 focal pigs to chew on a cotton bud for 1 min. Cortisol was analysed in duplicate by an enzyme immunoassay. Data were tested for normality, transformed and analysed in SAS (Proc Mixed) taking account of repeated measures (week). Acid-Buf™ reduced the number of aggressive (0.102 vs. 0.124 no./pig/min, s.e. 0.007; P<0.01) and mounting (0.005 vs. 0.014 no./pig/min s.e. 0.002; P<0.05) behaviours. Mounts were only performed by males and males also performed a higher frequency of aggressive behaviours than females (0.15 vs. 0.08 no./min/pig, s.e. 0.006; P<0.01). Acid-Buf™ pigs also tended to spend less time engaged in harmful behaviours than control pigs (18.6 vs. 23.9 secs, s.e. 2.11; P<0.06) and longer feeding (35.9 vs. 28.4 secs, s.e. 2.63; P<0.08). Finally, Acid-Buf™ pigs had lower salivary cortisol concentrations (1.30 vs. 1.47 s.e. 0.01 ng/ml; P<0.01). Pigs on a diet supplemented with Acid-Buf™ showed behavioural improvements leading to a calmer environment in which pigs could spend longer feeding and had lower stress levels.

## Conditioned place preference: a tool to determine hungry broiler diet preferences

*Buckley, Louise Anne[1], Sandilands, Vicky[1], Hocking, Paul[2], Tolkamp, Bert[1] and D'Eath, Rick[1], [1]Scottish Agricultural College, Kings Buildings, West Mains Road, Edinburgh EH9 3JG, United Kingdom, [2]Roslin Institute (University of Edinburgh), Easter Bush, Midlothian, EH25 9RG, United Kingdom; vicky.sandilands@sac.ac.uk*

The aim of the study was to determine if feed restricted broilers preferred quantitative feed restriction (QFR) or a diet containing the appetite suppressant calcium propionate (CAP) using a closed economy Conditioned Place Preference (CPP) task. Group-reared feed restricted broilers were allocated to one of two groups at 28 days and individually housed thereafter. The two groups were: QFR/AL (n=12) and QFR/CAP (n=12). Each bird alternated every two days between two diet options. For all birds one diet option was QFR. The other diet option was *ad libitum* (AL) feed access (QFR/AL) or QFR + CAP(QFR/CAP). The CAP inclusion rate was increased during the study (from 3-9%) to ensure maximal contrast in time taken to consume the diet. CPP training began on day 44. Birds alternated every two days between environments with horizontal or vertical black and white stripes. Each environment was paired with one of the diet options. Each bird was tested for a CPP after 12 days and 24 days with one test on a day when the bird had been fed QFR and the other on a day when the bird had been the other diet option. The test protocol included free-access to both pens for 20 minutes at the end of the day and the proportion of time spent in each pen was recorded. Differences from 0.5 were assessed using the One sample t-test. QFR/AL birds showed a state-dependent preference for the AL pen, spending more time in this pen only when hungry (t(11df) = 3.27, P=0.007; mean preference: 0.653, 95% CI: 0.550-0.757). QFR/CAP birds failed to show a significant preference for either pen, regardless of state. However, anecdotal observations suggested that the CAP option was not liked by the birds as they attempted to escape the pen on days when fed CAP. It is concluded that feed restricted broilers can learn a CPP task under certain circumstances. However, state-dependent expression of preference indicates that care is needed when using this methodology. The reason for the failure of QFR/CAP birds to express a CPP is unclear but includes a lack of preference or failure to learn the association.

**Feeding motivation and metabolites in pregnant ewes with different body condition scores**

Verbeek, Else[1,2,3], Waas, Joseph[3], Oliver, Mark[4], Mcleay, Lance[3], Ferguson, Drewe[1] and Matthews, Lindsay[2], [1]CSIRO, Livestock Industries, Locked bag 1, Armidale 2350 NSW, Australia, [2]AgResearch Ltd., Animal behaviour and Welfare, Private Bag 3123, Hamilton, New Zealand, [3]University of Waikato, Department of Biological Sciences, Private Bag 3105, Hamilton, New Zealand, [4]The Liggins Institute, Private Bag 92019, Auckland, New Zealand; else.verbeek@csiro.au

Long-term food restriction in pregnant ewes is likely to result in the subjective experience of hunger, which could affect welfare. We aimed to assess hunger by measuring feeding motivation and metabolic state in twin bearing ewes at different body condition scores (BCS, scored between 1 and 5). At 37 d of pregnancy, ewes were assigned to one of three groups to reach a low BCS (LBC, n=8), medium BCS (MBC, n=8) or high BCS (HBC, n=6) by 76 d of pregnancy. Blood samples were collected every 2 weeks from 37 to 133 d of pregnancy. Feeding motivation was assessed between 91 and 105 d of pregnancy using a behavioural demand methodology. The sheep were required to walk a set distance (costs: 2, 7.2, 13.8, 44 and 50 m) for a 5 g food reward over a 23 h period, with each cost tested on a separate day. The maximum price paid ($P_{max}$) and expenditure ($O_{max}$) were used as measures of motivation. Metabolic data were analysed using REML and correlations between $P_{max}$, $O_{max}$ and metabolic data were analysed using Spearman rank test. The average BCS were 2.0±0.0, 2.9±0.1 and 3.7±0.1 for LBC, MBC and HBC ewes, respectively,at the start of motivation testing.$O_{max}$ wassignificantly higher in LBC ewes (13.3 km) compared to HBC ewes (1.5 km), while the MBC ewes (9.2 km) were intermediate ($P<0.05$). No differences in $P_{max}$ were found. Glucose concentrations were higher in HBC ewes compared to MBC ($P<0.05$) and LBC ewes ($P<0.01$). Free fatty acid (FFA) concentrations were higher in LBC ewes than in MBC ($P<0.01$) and HBC ($P<0.001$) ewes. Glucose concentrations were negatively correlated to $P_{max}$ (r= -0.42, $P<0.05$) and $O_{max}$ (r=-0.49, $P<0.01$), while FFA was not correlated to $P_{max}$ or $O_{max}$. In conclusion, $O_{max}$ was inversely proportional to BCS, suggesting that feeding motivation is a potential indicator of the level of hunger experienced. Furthermore, as glucose concentrations reflected BCS and level of motivation, further investigation to validate glucose as an indicator of hunger is warranted.

**The relationship between aggression, feeding times and injuries in pregnant group-housed sows**

*Verdon, Megan and Hemsworth, Paul, Animal Welare Science Centre, Melbourne School of Land and Environment, University of Melbourne, Parkville, Vic, 3010, Australia; meganjverdon@gmail.com*

Concern for the welfare of stall-housed gestating sows has seen the phasing out of stalls in preference for group-housing. While group-housing allows increased movement and interactions, high levels of aggressionincreases injury and stress. Sows that avoid aggressive sows could also be sacrificing the opportunity to feed for safety. This study examined the relationships between aggression and feeding behaviour and injuries. 120 sows (parity 1-6) were randomly grouped within 7 days of insemination into pens of 10 with floor space of 1.4-3.0 m$^2$/sow. Feed was delivered via an over-header hopper onto the pen floor at 0700, 0800, 0900, 1000 each day and aggression (delivered and received) at feeding was recorded on days 2, 8 and 22 after mixing, with time spent feeding recorded on day 2. Sows were classified as 'Submissive' if they delivered no aggression, 'Subdominant' if they received more aggression than they delivered and 'Dominant' if they delivered more aggression than received. Both fresh (scratches, abrasions, cuts, abscesses) and old (partially healed) lesions were counted on days 2, 9 and 23 post-mixing. Relationships were examined using Kruskal-Wallis tests and repeated measures analysis of variance was used to examine the effects of time on aggression at feeding. Aggression significantly decreased over six 5-minute time intervals after a feed drop (means(+/-SE) of 2.04(0.057), 1.70(0.051), 1.32(0.034), 1.36(0.048), 1.35(0.053) and 1.18(0.050) aggressive interactions per sow, $P<0.001$), but didn't change between feed drops within days ($P=0.14$). The classification of sows (Submissive, Subdominant and Dominant) was associated with injuries (fresh lesions) on day 9 (means (+/-SE) of 16.31(2.02), 13.74(1.49) and 8.00(0.97) injuries per sow, $P<0.001$) and day 23 (means (+/-SE) of 23.63(9.83), 14.57(1.46) and 5.36(1.04), $P<0.001$). Furthermore, the classification of sows was associated with time feeding during the first (means(+/-SE) 0.33(0.052), 0.46(0.038) and 0.54(0.049) time feeding per sow, $P<0.001$) and second feed of the day (0.33(0.05), 0.48(0.032) and 0.45(0.039), $P=0.048$). The aggressive behaviour of sows appears to be associated with time feeding and injuries. In order to protect vulnerable sows in group-housing systems, more research into group dynamics and individual aggressiveness is required.

## Changes in aggression over time in pregnant sows post-mixing

*Rice, Maxine, Chow, Jennifer and Hemsworth, Paul Hamilton, Animal Welfare Science Centre, The University of Melbourne, Agriculture and Food Systems, Melbourne School of Land and Environment, The University of Melbourne, 3010, Parkville, Australia; mrice@unimelb.edu.au*

Increasing community concern about confinement housing has led to legislation, consumer and retailer pressure to increase the use of group housing for gestating sows. However, international industry experience indicates that the opportunity for group housing to improve sow welfare is presently limited by the high levels of aggression commonly observed in newly formed groups of sows after mixing. The present observations on sow aggression were conducted as part of a large project studying group-housing design for gestating sows. One hundred and twenty sows (parity 3-5) were randomly mixed into groups of 10 at 40 days post-insemination. Sows were housed in indoor pens with a floor space allowance of 1.6 m$^2$ per animal and fed 2.8 kg of a commercial diet delivered once a day (including at the time of mixing) via 2 drop feeders per pen. Using digital video records, observations on aggression (slashes, butts/pushes and bites) delivered and received were conducted for 60 min post mixing (Day 1) and 60 min post feed delivery on Days 2, 3, and 4. Kruskal-Wallis one-way ANOVA by ranks was used to examine the effects of time. There were significant effects of time on the number of aggressive behaviours delivered in the 1-h period following feed delivery (means (and standard error of the means) of 12.8, 11.5. 11.1 and 13.9 (SEM=1.36) aggressive behaviours per sow on Days 1, 2, 3 and 4 post-mixing, P=0.001). Conversely, there were also significant effects of time on the number of aggressive behaviours received in the 1-h period following feed delivery (means (and standard error of the means) of 12.6, 11.2, 10.7 and 13.5 (SEM=0.82) aggressive behaviours per sow on Days 1, 2, 3 and 4 post-mixing, P=0.017). Feeding and pen design features, such as feeding system, floor space and group size, are likely to affect aggression. The literature indicates that aggression should subside over time to low long-term levels by the 2$^{nd}$ or 3$^{rd}$ day post-mixing, however the relatively high levels of aggression around feeding over the first 4 days of grouping found in the present observations are surprising and highlight the need to identify practical strategies to reduce sow aggression.

## Feeding enrichment for Moloch Gibbons, *Hylobates moloch*

*Wells, Deborah, Irwin, Rosie, Hepper, Peter and Cooper, Tara, Queens University Belfast, School of Psychology, BT7 1NN, United Kingdom; d.wells@qub.ac.uk*

Feeding enrichment has been used successfully for a variety of primate species. Gibbons (*Hylobates* spp.), however, have been largely overlooked in relation to this type of enrichment strategy. This study therefore explored the effect of feeding enrichment on the moloch gibbon, a species that spends up to 70% of its time in the wild searching for, gathering and consuming food. A family group of four zoo-housed gibbons (2 adults, 2 offspring) were studied in response to three types of feeding device (food filled baskets, food filled PVC tubes, food frozen in ice pops). The animals were studied for 5 days during a control condition (i.e. no feeding enrichment) and 5 days per condition of feeding enrichment, when 3 of the same type of feeding device were suspended within the animals' exhibit. For each condition, the gibbons were studied using a scan-sampling technique every 5 minutes for 4 hours per day, providing 240 observations of each animal's behaviour per condition. The animals showed considerable interest in the feeders over the duration of their presentation, interacting with the devices an average of 24% of the available time. The subjects showed no significant difference (P>0.05, Friedman ANOVA) in the total number of times that they interacted with the 3 different feeding devices. There was no sign of habituation to any of the feeding devices over their five days of presentation (P>0.05, Friedman ANOVAs). Feeding enrichment significantly influenced certain components of the gibbons' behaviour, encouraging more species-typical patterns of activity (P<0.05, for all Friedman ANOVAs). Thus, animals spent more of the observation time outside (49.0% vs. 40.0%), showed more instances of foraging (18.1% vs. 8.4%), and fewer occurrences of moving (17.9% vs. 32.2%), during the enrichment conditions than the control. Overall, findings suggest that feeding devices may offer a valuable form of stimulation for captive-housed moloch gibbons, a species that has thus far been overlooked with regards this type of environmental enrichment.

### The effect of visitors on the behaviour of zoo-housed chimpanzees and gorillas

*Cooper, Tara C, Wells, Deborah L and Hepper, Peter G, Queens University Belfast, Psychology, University Road, Belfast, Northern Ireland, BT7 1NN, United Kingdom; tjenkins01@qub.ac.uk*

This study explored the effect of visitors on the behaviour of zoo-housed chimpanzees and gorillas. Two separate colonies of common chimpanzees (N=6) and western lowland gorillas (N=6) were studied in 2 conditions of visitor density. Condition 1 represented a period when visitors were devoid from the viewing areas of the animals' exhibits; Condition 2 represented a period when 1-20 visitors were present in the viewing areas. Animals were tested between April-November and there was little significant difference in weather at the time of the observations between conditions. Every animal in the two groups was observed, using focal sampling, for 12 ten-minute periods, between 12.00 noon and 2.00 pm, in each condition, providing 720 minutes of data per animal per condition. Analysis revealed a significant effect of condition on the behaviour of both groups of apes. Animals spent significantly ($P<0.05$, ANOVAs) more time running and walking in Condition 2 (mean duration of observations: running=2.14±(SD)2.67, walking=11.51±5.27) than Condition 1 (running=0.22±0.66, walking=6.80±5.48). Both groups spent significantly ($P<0.05$, ANOVAs) less time displaying inactive (e.g. sitting, lying down) and visitor-oriented behaviours in the presence (inactive=84.28±7.82, visitor orientation=10.71±9.03) than absence (inactive=92.16±5.55, visitor orientation=0±0) of visitors. Significant interactions ($P<0.05$, ANOVA) between condition and group illustrated behavioural differences between the species. Chimpanzees foraged for significantly longer in the presence (5.40±3.62) than absence of visitors (2.46±1.62); the reverse pattern was true for gorillas (visitors present=8.13±6.76, visitors absent=14.35±5.65). Likewise, chimpanzees spent significantly less ($P<0.05$, ANOVAs) time allo-grooming and displaying self-directive (e.g. anal stimulation) patterns of behaviour in the presence (allo-grooming=4.52±5.78, self-directed=7.19±5.35) than absence (allo-grooming=19.70±13.02, self-directed=14.45±9.38) of visitors; gorillas, by contrast, showed higher incidences of self-directive behaviour in the presence (8.83±14.38) than absence (8.45±10.33) of the human audience. The results suggest that whilst the behavioural response of chimpanzees to visitors suggests that they find them enriching the behavioural response of gorillas suggests that they find visitors stressful.

## Olfactory enrichment in the gorilla

*Hepper, Peter, Wells, Deborah and Jackson, Rachel, Queens University Belfast, Psychology, Belfast, BT71NN, United Kingdom; p.hepper@qub.ac.uk*

Although apes respond to olfactory signals, the use of odor stimulation as an enrichment tool in such animals has had little success. This raises the question of whether the lack of effect is due to a methodological issue. Previous studies have exposed olfactory stimuli to animals on cloths. This study compared 2 modes of odor presentation, namely, a cloth or a patch sprayed in the enclosure. Three odor stimuli (banana, vanillin, water) were used. A group of six lowland gorillas (3 males, 3 females), housed at Belfast Zoo, were studied. Odors were presented sprayed either: on a 70×70 cm flannel cloth (n=6) or, as a 70×70 cm square patch (n=6) on the grass of the outdoor enclosure. Animals were studied over a 5 day period in each condition, e.g. banana on cloth. Fresh stimuli were presented on day 1, 3 and 5 and the animals' behaviour recorded on these days. Cloths were removed at the end of day 1 and 3 and no detectable odor was present from the patches at day's end. Animals were observed (scan-sampling every 5 minutes) from 0900-1200 each day. Two behaviours were recorded: sniffing, the animal bought the cloth within approximately 10 cm of its nose, or placed its nose within 10 cm of the cloth, patch; locomotion, the animal was observed moving. An analysis of variance examined method of presentation (cloth, patch) and odor (vanillin, banana, water) for sniffing and locomotion separately. There was a significant odor x presentation interaction (P<0.05) for sniffing and locomotion. There was no difference in the percentage number of sniffs exhibited by gorillas when water was used (cloth 3.40[+/-s.d. 1.18] vs patch 3.70[1.01]) but more sniffs were observed with odors when patches were used compared to cloths (vanillin 10.19[2.28]) vs 7.87[2.03]), banana 12.50[2.15]) vs 8.64[2.03]). There was no difference in locomotion when water was used (cloth 13.73[3.21], patch 14.97 [3.06]) but with odors, gorillas moved significantly more when patches were used compared to cloths (vanillin 21.45[3.62] vs 16.05[3.07], banana 22.38[3.45] vs 15.12[2.90]). Presenting odors as patches in the animals' environment elicited a greater response than when presented on cloths. We suggest that patches are a more ecologically valid means of odor presentation than artificial objects and the use of olfactory stimulation as an enrichment tool in apes may be more effective if presented naturalistically.

## Exotic animal seasonal acclimatization determined by non-invasive measurements of coat insulation

*Langman, Vaughan, Langman, Sarah, Ellfrit, Nancy and Soland, Tine, United States Department of Agriculture, Animal Care, 816 White Pine Drive, 80512 Bellvue, Colorado, USA; macora@q.com*

Seasonal acclimatization in terrestrial mammals in the Northern Hemisphere involves changes in coat insulation. This study was done to test a technique for the non-invasive measurement of mammal coat insulation and to measure coat insulation over several seasons on captive exotics. The working hypothesis was that species that have no coat or have a coat that does not change seasonally do not acclimatize seasonally. Insulation was measured in the winter and summer of 2001, 2002 and 2003 at Cheyenne Mountain Zoo in Colorado Springs, Colorado. The animals measured were three Amur tigers, two African elephants, ten giraffe, one okapi, and three mountain sheep. To calculate the insulation value of the animal's hair coat it is necessary to measure the heat escaping from the animal's body. Insulation was measured non-invasively on all of the animals in this study in light wind and out of direct solar radiation using infrared guns and thermal imagers to measure the surface temperature of the coat or skin. Three surface temperature readings were measured from the torso area. The insulation was calculated using measured metabolic rates and body temperature when possible. The African elephants, giraffe and okapi did not acclimatize and maintained insulation values year around for an ambient temperature of 21 °C. The Amur tigers and mountain sheep acclimatized to seasonal ambient conditions by increasing the insulation values of the hair coats in the cold and loosing the hair coat in the summer to decrease their insulation values. The Amur tigers and mountain sheep were insulated over a range of ambient air temperatures from -10 °C to a +35 °C. These species had insulation values under cold conditions three to four times greater than the insulation values measured for African elephants, giraffe and okapi under the same conditions. The husbandry implications of exotics that have no ability to acclimatize to Northern Hemisphere seasonal ambient air temperature changes are profound. Giraffe, African elephants and okapi when exposed to cold conditions with ambient air temperatures below 21 °C will use body energy reserves to maintain a heat balance and require an increase in the quality and quantity of food during cold periods. All of these species will require housing that provides ambient conditions of 21 °C.

**Effect of social rank and lactation number on milk parameters in a grazing dairy herd**
*Tresoldi, Grazyne[1], P. Machado F⁰, L. Carlos[1], Sousa, Rafaela C.[1], Rosas, M. Inés[2] and Ungerfeld, Rodolfo [2], [1]LETA – Lab. de Etologia Aplicada, Depto. de Zootecnia e Des. Rural, Universidade Federal de Santa Catarina, Rod. Admar Gonzaga 1346 – Itacorubi, 88034-001, Florianópolis, SC, Brazil, [2]Facultad de Veterinaria, Universidad de la Republica, Alberto Lasplaces, 1550, Montevideo, Uruguay; pinheiro@cca.ufsc.br*

Dominant cows have priority access to resources, wich may result in a better nutrition. However, fear to humans may increase milk retention and affect their udder health. We studied the effect of social order and lactation number on milk parameters in a herd of 102 Holstein grazing cows. A total of 32,000 observations of agonistic interactions were recorded in 3 months. A sociometric matrix was calculated and cows were assigned to a rank group (D-Dominants n=23; I-Intermediates n=47; S-Subordinates n=32) according to their individual dominance value. They were also classified according to their lactation number in L1-first lactation; L2-second lactation; L3-third lactation; and L4-fourth lactation and higher. Daily milk production and milking time were recorded for each individual animal and the milk flow (ml/s) was calculated. Likewise, somatic cells count (SCC) and mastitis occurrence were evaluated. Data was statistically evaluated using ANOVA. A regression of social rank on lactation number was calculated. Daily milk production from I was higher than from S (I=27.0 ±0.9, S=23.7 ±1.1; P=0.02), without differences from D (25.5 ±1.3) cows. Milking was longer for S (445.9 s ±17.3) than for D (398.1 s ±20.4, P=0.07) and for I (389.1 s ±14.3; P=0.01) cows. Milk flow was higher in I than in S (I=71.3 ml/s ±2.9, S=56.7 ml/s ±3.5; P=0.002), without differences from D (65.4 ml/s ±4.1) cows. Daily milk production increased with lactation number, although with no difference between L2 and L3, and L3 and L4 (L1=21.7 ±1.1, L2=25.3 ±1.1, L3=27.1 ±1.2, L4=29.0 ±1.2; P=0.03). Milk flow was lower (P=0.004) in L1 than in L2, L3 and L4 cows (L1=52.0 ml/s ±3.6, L2=67.1 ml/s ±3.6, L3=75.0 ml/s ±4.0, L4=69.6 ml/s ±3.8). No differences were found among groups for SCC (192.598 ±33.9) and for mastitis occurrence (1.3 ±0.1). A positive and significant correlation between social rank and lactation number (r²=0.46, P<0.0001) was found, what may explain the lower milk production and milking flow for subordinate cows in a large, heterogeneous dairy herd as the studied one.

## Does walking through water stimulate cows to eliminate?

*Villettaz Robichaud, Marianne[1], De Passillé, Anne Marie[2] and Rushen, Jeffrey[2], [1]Université Laval, Sciences Animales, Faculté des sciences de l'agriculture et de l'alimentation, Pavillon Paul-Comtois, Québec, G1V 0A6, Québec, Canada, [2]Agriculture and Agri-Food Canada, Pacific Reseach Centre, 6947, Highway 7, Cp. 1000, Agassiz, V0M 1A0, BC, Canada; marianne.villettaz@gmail.com*

Manure is a source of disease and health problems in humans and cows. The mix of feces and urine releases volatile ammonia, an important greenhouse gas. Keeping feces and urine separate would reduce the environmental impact of dairy production. We tested ways to stimulate 12 lactating Holstein cows (Day in milk= 137.5±17.5 d, parity = 3.3±1.5) to defecate or urinate. In Test 1, cows were walked through an empty or a water filled footbath (20 °C) following a balanced order, with one treatment/d, over 6 d. Each cow was exposed 3 times to each treatment. Cows were more likely to defecate with the footbath filled with water (24/36) then without water (16/36) (Sign test: P=0.04). In Test 2, the cows stood 2 mins in an empty footbath, or in a footbath filled with either still or running water, with one treatment/d, over 9 d. In Test 3, the cows stood for 2 mins in an empty footbath with nothing, air or water sprayed on their feet, with one treatment/d, over 9 d. No treatment differences were found for Tests 2 and 3 (Sign test: P>0.10). After Test 3, we repeated Test 1 but no treatment difference was found at that point (Sign test: P>0.10). The frequency of defecation decreased from an average of 50% on day 1-2 to an average of 21% on day 25-26 of the experiment. Contrary to what is sometimes claimed, water does not seem to stimulate cows to defecate or urinate in a systematic way. Defecation and urination might decrease with gradual habituation of cows.

## Factors predicting horse welfare outcomes from a recreational horse owner's performance of key horse husbandry practices

*Hemsworth, Lauren M[1], Jongman, Ellen[2] and Coleman, Grahame J[1], [1]Monash University, Clayton, 3800, Australia, [2]The Department of Primary Industries, Werribee, 3030, Australia; Lauren.Hemsworth@monash.edu.au*

The welfare of recreational horses in Victoria, Australia is an increasingly important issue. A substantial proportion of horse welfare problems appear due to horse owner mismanagement, resulting from ignorance rather than intentional abuse. According to the Theory of Planned Behaviour, a horse owner's attitudes towards horse ownership may influence behaviour in terms of the implementation of horse husbandry practices. Subsequently these behaviours may impact on the welfare of the horse. This was an observation-based investigation into the antecedents of Victorian recreational horse owner husbandry behaviour, and the ensuing relationship with horse welfare outcomes. Horse owner attribute data and horse welfare outcome data were collected during inspections of 57 Victorian recreational horse owners and their horses. Bivariate correlation and multiple regression analyses examined the performance of three husbandry practices: parasite control, hoof care and dental care. Rural primary residence, younger age and regular riding instruction were significantly correlated with positive salient beliefs for all three husbandry behaviours (P<0.05). A horse owner's membership to a horse club/society was significantly correlated to positive salient beliefs regarding hoof care and dental care behaviour (P<0.05). Regarding parasite control and dental care, positive 'attitude towards behaviour' and 'perceived behavioural control' accounted for significant variance in the performance of appropriate husbandry behaviour ($R^2$=0.34 and $R^2$=0.26 respectively). Positive 'attitude towards behaviour' accounted for significant variance in appropriate hoof care ($R^2$=0.42 P<0.01). The inappropriate performance of all three husbandry behaviours were significantly correlated with poor horse welfare outcomes, based on body condition, hoof condition, lameness and injury scores (P<0.05). Clearly, relationships exist between Victorian recreational horse owner attributes, husbandry behaviours and subsequent horse welfare outcomes. Understanding the nature of these relationships is crucial in order to develop and implement appropriate strategies to manage the welfare of recreational horses in Victoria.

**Moderate exercise affects finishing cattle behavior and cortisol response to handling**
*Glynn, Hayley, Stickel, Andrew, Edwards, Lily, Drouillard, Jim, Houser, Terry, Rozell, Tim, Jaeger,
John, Hollis, Larry, Miller, Kevin and Van Bibber, Cadra, Kansas State University, Department of
Animal Sciences and Industry, Manhattan, KS 66506, USA; hdglynn@k-state.edu*

The objective of this study was to determine the behavioral and physiological changes associated with exercise in finishing cattle, providing preliminary data to assess the potential application of routine exercise in feedlots as a means to improve cattle welfare. Thirty cross-bred heifers (n=15; 448±27 kg) were stratified by weight and allocated randomly to sedentary (CON) or exercise (EX) treatments. All cattle were housed individually (1.5 m X 6.5 m pens). Each Mon, Wed, and Fri morning, EX cattle were removed from their pens and moved by an animal caretaker at a pace of 5 to 6 km/h (20 min/d for the first 2 wk, 30 min/d for the next 2 wk, and 40 min/d thereafter) for an 8 wk period. Blood was sampled via jugular venipuncture and analyzed for cortisol on day 0, 28 & 60. Behavior in pens prior to exercise sessions (n =11) was videotaped during weeks 5, 6 & 7 on 2 of the 3 exercise days (EX) and 2 corresponding days for the CON group per week. Video was viewed using instantaneous sampling with a 30-s interval. Cortisol concentrations and percentages of time cattle spent with head in bunk, standing, lying and moving were analyzed as repeated measures using mixed models with fixed effects of treatment, week and the interaction, with appropriate random effects. A treatment*day interaction effect was observed for cortisol (P<0.01); EX cortisol was lower on d 28 & 60 than d 0 (P<0.01), but concentrations in CON cattle were unaffected by time (P ≥ 0.21). Cortisol tended to be greater for CON than for EX on d 60 (P=0.06; 48.0±4.3 and 36.5±4.3 ng/ml, respectively). There was a treatment*week interaction effect for time spent moving (P=0.04), and tendencies for effects on time spent with head in the bunk, lying, and standing (P ≤ 0.07). EX cattle moved more in their pens than CON cattle during weeks 5 & 6 (P ≤ 0.04). EX cattle spent less time moving during week 7 than week 5 (P=0.04; 1.3±0.7 & 3.6±0.7%, respectively). Distinct behavioral patterns were not identifiable during the 3 week period. The reduction in cortisol measurements in the EX group suggest that cattle became accustomed to handling via routine exercise thus reducing the stress response during subsequent processing.

**Humans' perception of dogs in a research setting: Is there a difference between real dogs and virtual dogs?**

*Stetina, Birgit U.[1], Kastenhofer, Elisabeth[2], Hauk, Nathalie[2], Glenk, Lisa M.[3] and Kothgassner, Oswald D.[2], [1]Webster University, Department of Psychology, Berchtoldgasse 1, 1220 Vienna, Austria, [2]University of Vienna, Clinical Psychology, Liebiggasse 5/3, 1010 Vienna, Austria, [3]Karl Landsteiner Research Institute for Pain Treatment and Neurorehabilitation, Neurochemical Laboratory, LKM Mauer-Amstetten, 3300 Amstetten, Austria; stetina@webster.ac.at*

Introduction: The way humans perceive animals has a crucial impact on human-animal interaction research. This pilot study addressed differences in human emotional status and perception by comparing the presence of a real animal versus a virtual animal. Methods: The 110 participants (mean age 27.5 years) were randomly assigned to two test groups and one control group. Test group 1 was exposed to real dogs. In test group 2, the participants were exposed to virtual simulations of the same dogs, seen through Head-Mounted-Displays. In order to cover a variety of dog types, two specially trained animals partook in the pilot study: a Flat-Coated Retriever and a Rhodesian Ridgeback. Two eight-minute exposure times with a two minute break were utilized. The German version of PANAS was implemented to measure the humans' emotional status. An especially developed questionnaire on the subjective perception of dogs' behaviours and corresponding attributions was used. Results: Both test groups showed significant increases in positive emotional affects (such as happiness or serenity) after the short time exposition in comparison to the control group (F (2, 78)=2.002; P=0.048). There were no significant differences between the control and the test groups in negative emotional affects (i.e. sadness, fear) after the exposition, (F (2, 78)=1.002; P=0.398). More specifically, no significant effects were found between the experimental groups (real and virtual dog). In addition, only 3 of 20 attributions showed significant differences between the real and virtual dogs. The real dogs were perceived as significantly more affectionate (P=0.040, d=0.68), friendly (P=0.021, d=0.76) and trusting (P=0.014, d=0.85). Discussion: The impact on emotional status and perception, using dogs in virtual reality compared to real dogs, seems to be relatively similar. Especially startlingly was the analogous perception of different attributes of the virtual and real dogs.

**Finding information on animal behavior and welfare: best practices for starting the literature review process**

*Adams, Kristina, US Department of Agriculture, Agricultural Research Service, National Agricultural Library, Animal Welfare Information Center, 2150 Centre Avenue, Building D, Ft. Collins, CO 80526, USA; kristina.adams@ars.usda.gov*

A critical step in any research project is conducting a comprehensive review of the literature in order to identify current knowledge as well as potential gaps about a particular topic. Graduate students and seasoned researchers in applied behavior science use the literature review process to identify relevant case studies on a topic and learn about new or existing methodologies when defining a theoretical framework for their own studies. A major step in the literature review process is conducting a comprehensive search of published literature, however, training in information seeking behavior and knowledge of library and online resources is often lacking. This paper will provide tips for conducting a literature search, including where to go, how to search, and how to manage citations. The discussion will include an overview of bibliographic databases that index animal behavior and welfare information and demonstrate search techniques useful for finding relevant papers. Along with the more traditional searching tools and techniques, newer time-saving techniques will also be presented. With the ever increasing amount of literature available online, scientists possessing the knowledge about where to go for information and the skills to conduct a thorough search will ensure a more complete review of their area of study.

**An assessment of the general activity of horses kept in large groups in a feedlot environment**

*Robertshaw, Marissa[1,2,3,4], Pajor, Ed[5], Keeling, Linda[6], Burwash, Les[7] and Haley, Derek[1,2,3,4], [1]Ontario Veterinary College, University of Guelph, Population Medicine, 2538 Stewart Building, Guelph, ON, N1G 2W1, Canada, [2]Ontario Veterinary College, University of Guelph, Population Medicine, 2538 Stewart Building, Guelph, ON, N1G 2W1, Canada, [3]Ontario Veterinary College, University of Guelph, Population Medicine, 2538 Stewart Building, Guelph, ON, N1G 2W1, Canada, [4]Ontario Veterinary College, University of Guelph, Population Medicine, 2538 Stewart Building, Guelph, ON, N1G 2W1, Canada, [5]Faculty of Veterinary Medicine, University of Calgary, Production Animal Health, HSC 2517, Calgary, AB, T2N 1N4, Canada, [6]Swedish University of Agricultural Sciences, Animal Welfare, Box 7068, Klinikcentrum, Travvägen 10D, 750 07 Uppsala, Sweden, [7]Alberta Agriculture & Rural Development, Livestock Business Development Branch, 97 East Lake Ramp NE, Airdrie, AB, T4A 2G6, Canada; mrober11@uoguelph.ca*

In western Canada, keeping horses outdoors in large groups (e.g., 100 to 200 animals) in pens with dirt flooring is a common method of management when feeding horses for meat production. We have found no published scientific studies documenting the behaviour of horses under these conditions. Because of their social evolutionary history and natural group sizes some have expressed concern that horses may not be well-suited for life in a feedlot environment. As a first step in considering the possible welfare implications of keeping horses in this manner, we documented the general activity of 2 pens of horses containing 162.00±9.90 and 176.30±20.30 (mean±SD) individuals on observation days, housed at stocking densities of 41.25±68.25 and 25.38±36.25 m$^2$, respectively. Over a period of 2.5 months, beginning in June 2010, we recorded activity by live observation, two days each week (a total of 144 h of observation/pen). Observations were made over a total of 8 h/d, in 4, 2-h blocks: from 0700 to 0900 h, 1000 to 1200 h, 1300 to 1500 1600 to 1800 h. At 10-min intervals, the number of horses in each pen engaged in each of the following behaviour patterns was counted: standing, lying, eating, walking, grooming, playing, and 'other'. At any one point in time 46% of horses were standing idle, 26% eating hay or concentrate, 12% lying down, 6% walking, 5% grooming, 1% playing and 4% other. Further details from this descriptive study, will be provided on the final poster.

## The Animal Welfare Juddging & Assessment Competition: a review of the 1st 10 years

*Heleski, Camie[1], Golab, Gail[2], Millman, Suzanne[3], Reynnells, Richard[4], Siegford, Janice[1] and Swanson, Janice[1], [1]Michigan State University, East Lansing, MI 48824, USA, [2]American Veterinary Medical Association, Schaumburg, IL 60173, USA, [3]Iowa State University, Ames, Iowa 50011, USA, [4]United States Department of Agriculture, Washington, DC 20250, USA; heleski@msu.edu*

In 2001 Heleski, Zanella & Pajor presented to ISAE the idea of promoting animal welfare science to university students by coupling it with the more traditional concept of livestock judging. In 2002 MSU hosted the first AWJAC for 4 teams representing 4 universities, 18 participants. In 2010 we hosted the 10th AWJAC for 18 teams representing 9 universities, 78 participants. Originally the contest was for undergraduates but now consists of 3 divisions: undergraduate, graduate and veterinary students. Initially the AWJAC was conducted only with livestock species; now it represents production, companion, laboratory and exotic animals. The AWJAC relies on hypothetical, realistic computer-viewed scenarios containing performance, health, physiologic & behavior data. These are evaluated individually; students determine placings as to which facility has a higher welfare level and present oral reasons regarding their rationale. Reasons are presented to judges with expertise in animal welfare science and specific knowledge of the species they are judging. Students also participate in a team assessment exercise, usually conducted live at an animal facility. Teams of 3-5 students give group presentations to a panel of judges; this might consist of their recommendations for welfare-related changes at the facility. Students are surveyed at the end of each AWJAC to obtain perceptions about the AWJAC. Over 95% of participants believe the AWJAC is a valuable exercise, feel they have increased their knowledge about welfare science & would recommend the AWJAC to peers (n=345). In response to student feedback, in 2006 we switched to a 2 day format and added a speaker program. This educational component has been extremely well received. While the assessment of various aspects of animal welfare can be objective & quantifiable, ethics-based choices may impact where on the welfare continuum is considered acceptable, preferred or unacceptable. The AWJAC teaches students to integrate science-based knowledge with ethical values for an interdisciplinary approach to problem solving.

**Differences in the behavior of the progeny of different sire lines of Japanese Black cattle**

Uetake, Katsuji[1], Ishiwata, Toshie[1], Kilgour, Robert[2] and Tanaka, Toshio[1], [1]Azabu University, School of Veterinary Medicine, 1-17-71 Fuchinobe, Chuo-ku, Sagamihara, 252-5201, Japan, [2]Industry and Investment New South Wales, Agricultural Research Centre, Trangie, NSW, 1823, Australia; uetake@azabu-u.ac.jp

We determined differences in the behavior of the progeny of two major sire lines of Japanese Black cattle by recording the behavior of 35 and 70 half-sib steers of sires from fast (FG) and slow (SG) growing lines. Two sire lines of steers were mixed and allocated to nine pens with 11-12 animals per pen. Behavior of the steers was recorded using instantaneous scans made at 15-minute intervals during the hours of daylight (05.00-18:00 hours) for four days. The proportion of steers lying was significantly ($P<0.001$) higher in the SG line ($43.4\pm5.7\%$ compared to $40.3\pm6.0\%$). The proportion of time spent eating concentrate feed (FG: $12.1\pm2.3\%$; SG: $11.4\pm2.1\%$), drinking (FG: $0.8\pm1.1\%$; SG: $0.4\pm0.6\%$), licking the feed trough (FG: $0.4\pm0.6\%$; SG: $0.2\pm0.4\%$) and performing tongue-playing (FG: $3.1\pm4.6\%$; SG: $1.0\pm1.9\%$) was significantly higher in FG, whereas the proportion of time spent resting (FG: $41.5\pm12.8\%$; SG: $43.7\pm10.9\%$) and performing self-licking (FG: $1.7\pm1.4\%$; SG: $2.1\pm1.3\%$) was higher in SG (all $P<0.05$). These results show progeny of the fast-growing sire engaged in more active behaviors compared to the progeny of the slow- growing sire line.

### Difference in pawedness between male and female blue foxes (*Vulpes lagopus*)

*Mononen, Jaakko[1], Tikka, Sanna[1,2] and Korhonen, Hannu T.[2], [1]University of Eastern Finland, Department of Biosciences, P.O. Box 1627, 70211 Kuopio, Finland, [2]MTT Agrifood Research Finland, Animal Production Research, Regions, P.O. Box 44, 69101 Kannus, Finland; jaakko.mononen@uef.fi*

It has been hypothesized that stress may affect brain and behaviour lateralization, and therefore lateralization tests might be useful in animal welfare research. We developed a pawedness test for farmed blue foxes, and used the test for studying the effects of sex on the paw preference. The blue foxes (78 female, 35 male; age 10-34 months) were housed singly, and the tests were performed in their home cages. In the test a titbit was attached outside the front wall of the wire mesh cage to persuade a fox to use its forepaws to reach for the titbit. The observer recorded whether a fox used first its left (L) or right (R) forepaw for reaching the titbit. The test was repeated for each fox once a day on ten days, and a laterality index was calculated: $LI=(R-L)/(L+R)$, where L and R are the numbers of times each paw was used first. Deviation of LI from zero was tested separately for the two sexes with the Wilcoxon's Signed Ranks Test. The effect of sex on LI was compared with Mann-Whitney U-test. Fisher's Exact Test was used to compare the number of females and males showing laterality at individual level (LIND), defined as an animal using the same paw first in 9 or 10 tests (i.e. $P<0.05$, Binomial Test). The paw test was easy to carry out, and the foxes habituated quickly to the test situation. In the end of the test period the testing of the 113 animals took less than 3 hours per day. LI was biased to the left in the females (quartiles: -0.6, -0.2, 0.2; $P<0.05$; n=78), whereas no difference from zero was observed in the males (quartiles: -0.4, 0, 0.6; $P=0.67$; n=35). LI did not differ between the sexes ($P=0.11$), but the number of LIND animals was biased more ($P<0.05$) to the left in the females (L=17, R=7) than in the males (L=3, R=8). The percentage of LIND animals was 31% in both sexes ($P=1.0$). We conclude that pawedness of farmed blue foxes can be measured with a simple and feasible test. Several studies have shown bias to the right in the female and to the left in the male dogs' (Canis familiaris) forepaw use, but in V. lagopus, ie. another canid species, the situation seems to be rather the opposite. This finding suggests a need for wariness in making any generalizations on the effects of sex on behavioural laterality.

## Effect of an absence of wires from electric fences on movement of goats in the early stage of avoidance

Kakihara, Hidetoshi[1], Ishiwaka, Reiko[2], Masuda, Yasuhisa[2], Nakano, Yutaka[1], Izumi, Kiyotaka[1], Horie, Chihiro[1], Furusawa, Hirotoshi[1] and Shimojo, Masataka[1], [1]Kyushu University, Fukuoka, 812-8581, Japan, [2]Kuju Grassland Ecomuseum, Fukuoka, 812-0053, Japan; 3BE10009R@s.kyushu-u.ac.jp

Electric fences are applied in many cases where ruminants such as cattle, sheep and goats are grazing because the fences require considerably less time and materials to construct. Many scientists have so far considered that animals can be trained to avoid electric fences by receiving a shock. The objective of this study is to know from what component of fences animals flee in avoidance of the fences through an investigation on the effect of an absence of wires on behaviour and position of grazing goats in the early stage after initial learning. Four Tokara dwarf nanny goats were trained for an electric fence and released separately to a paddock fenced with a single electrified wire. Colleagues of each target individual were left beyond the wire. Goats were observed for 20 minutes directly after separation from her colleagues in two different conditions within a day. One condition was with an electrified wire stretched across the paddock (control), whereas the other was with the wire stretched a half length on the same line, another half left opened (partly non-wired). We recorded behaviour and position of grazing goats at a ten-second interval and compared them between conditions. The electric fence was removed after each observation. No animal got out of the paddock in both control and partly-wired condition and some individuals made a right-about-face immediately in front of the fence even if a half of the wire had been removed. The minimal distances between an individual and the electric fence line in partly non-wired condition were significantly less than those in control (partly non-wired; 3.1±2.7 m vs. control; 4.7±3.1 m, P<0.05, paired t-test). Learning that the fence is an object to be avoided usually occurs immediately after receiving a shock. Our results show that wires are not necessarily included in the perception of the fence; goats may flee from the other components such as poles and/or area than wires of the fence. Grazing goats may look on an electric fence as a wall at least in the early stage after initial learning.

### Beef cattle preferences for sprinklers

*Parola, Fabia[1,2], Hillmann, Edna[2], Schütz, Karin E[3] and Tucker, Cassandra B[1], [1]UC Davis, Davis, CA, USA, [2]ETH, 8092, Zürich, Switzerland, [3]AgResearch, Ltd, Hamilton, New Zealand; cbtucker@ucdavis.edu*

Sprinklers effectively reduce heat load in cattle. In some studies, cattle readily use sprinklers, while others find that they either avoid or show no preference for it. These studies differ in many ways including previous experience of the animals and design of the sprinkling systems. We examined preferences for a system with nozzles mounted over the feed bunk that intermittently spray (on for 7 min, off for 13 min/24 h in this experiment) the animals' back while they eat. This system is common on US dairies, but we used beef cattle in this work because they had no experience with man-made water cooling. Steers (366±50 kg) were housed in groups of 4 (n=8 groups; 4 groups tested simultaneously) and provided 2 unshaded feed bunks in adjacent, connected pens with and without sprinklers in summer. Each group was tested with 1 of 2 nozzle types at a time, delivering either 1.3 or 2.6 l/min, thus at any given time, animals chose between pens with sprinklers of 1 nozzle type and or no sprinklers at all. To ensure steers were making an informed choice, they were first fed solely from each feed bunk (with and without sprinklers) for 1 d/each per nozzle type. Preference was then evaluated for 3 d from 13:00 to 19:00 (32.5±4.8 °C during this time) for all groups by feeding the animals *ad libitum* from both feed bunks and recording their behavior with 10-min scan sampling. The percentage time spent in the sprinkler treatment was compared to 50% with a 1-sample t-test. Steers tended to spend more time within the sprayed area in the 1.3 l/min treatment (62% of 6 h, P=0.06), but not when the 2.6 l/min nozzles were used (56% of 6 h, P=0.22). On average, steers showed no preference for the sprayed area near the feed bunk when the water was on (51 and 42% of 6 h for 1.3 and 2.6 l/min, P≥0.24), nor did they prefer to stand with their head in the feed bunk when water was on or overall (overall: 55 and 57% of 6 h for 1.3 and 2.6 l/min, P≥0.16). However, this pattern was influenced by weather. Steers spent more time near the feed bunk with sprinklers (both nozzle types) as it became warmer (P<0.01; 36% of 6 h at 25 °C to ≥71% of 6 h at 41 °C). In conclusion, naïve beef cattle prefer direct spray from sprinklers fitted with nozzles that deliver 1.3 or 2.6 l/min, but this preference is dependent on ambient temperature.

**Correlations between fearfulness in a temperament test and activity levels in the home pen in cross bred beef steers**

*Mackay, Jill R D[1,2], Turner, Simon P[1], Hyslop, Jimmy[1] and Haskell, Marie J[1], [1]SAC, West Mains Road, Edinburgh, EH9 3JG, United Kingdom, [2]University of Edinburgh, Ashworth Laboratories, Kings Buildings, West Mains Road, Edinburgh, EH9 3JT, United Kingdom; jill.mackay@sac.ac.uk*

Tri-axial activity monitors have been used to remotely characterise the behaviour of large groups of animals. These tools may be useful for welfare assessment if they can detect patterns in activity which relate to behavioural traits. This study investigated one possible use of the Motion Index™ (MI), an automatic measure derived from IceTags (IceRobotics Ltd., South Queensferry, UK). The MI is a measure of the absolute acceleration of the limb the tag is attached to. In beef cattle, two frequently used temperament tests are the Crush Score (CS) and Flight Speed from a crush (FS), which are commonly considered to reflect the underlying personality trait of fearfulness. How these traits relate to behaviour in the home pen is so far unknown. Thirty eight Aberdeen Angus cross (AAX) and 30 Limousin cross (LMX) beef steers were kept in four groups for 57 days. The animals were weighed weekly and on Days 15, 29, 43 & 54 of the trial they were CS and FS tested. MI data were collected by IceTag Pros fitted to the left hind leg from Days 29-42 (13 days) for half the animals and from Days 15-28 & 43-53 (26 Days) for the other half. MI was calculated using IceTagAnalyser 2010 at daily intervals and then averaged for each animal (DMI). The CS and FS showed good concordance between test days with a Kendall's coefficient of 0.481 & 0.669 respectively. FS did correlate weakly with CS explaining 12.8% of the variance (P=0.002). CS did not correlate with DMI for AAX or LMX (P=0.385, P=0.701). For LMX animals, FS did not correlate with DMI (P=0.432). However, for AAX animals DMI explained 25.7% of the variation in FS (P<0.001). The results of this study suggest that the underlying trait which drives variation in FS also affects the behaviour of AAX animals within the home pen and that activity monitors could be an alternative to certain temperament tests. But this relationship does not hold true for the LMX steers. Overall we conclude that activity monitors are a useful tool for understanding how temperament affects behaviour in the home pen, and can be used to more fully explain what such temperament tests might be measuring.

**Behavioural differences in Nelore cows following use of an progesterone-releasing intravaginal device**

*Rueda, Paola Moretti[1,2], Lima, Victor Abreu de[2], Sant'anna, Aline Cristina[1,2] and Paranhos Da Costa, Mateus José Rodrigues [1,2], [1]Faculdade de Ciências Agrárias e Veterinárias, Unesp, Jaboticabal, 14884-900, Brazil, [2]Grupo de estudos em etologia e ecologia animal, Grupo ETCO, Faculdade de Ciências Agrárias e Veterinárias, UNESP, Jaboticabal, 14884-900, Brazil; paolamrueda@yahoo.com.br*

The use of progesterone-releasing intravaginal devices (PRID) is growing due to its application for fixed-time artificial insemination and embryos transfer techniques, but the effect of this device on cow comfort remains unclear. The aim of this study was to assess the effect of PRID on the behavior beef cows, testing the hypothesis that the device has a negative effect on cow welfare. The study was carried out with 35 Nellore multiparous cows (without calves), which were habituated to the corral and handling procedures. The PRID insertion was done after restraining each cow in a squeeze chute. Behavioral data were recorded for 30 minutes over six consecutive days, three of them before and three after the device insertion. The cows were kept on pasture, and driven once a day to the corral, maintaining them in a 420 $m^2$ corral pen with free access to water from 9:00 to 10:00 h, when the observations were done. Four behavioral categories were recorded, two postures (standing – ST or lying – LY) and two activities ruminating (RU) or idling (ID). Data were recorded using instantaneous record (one minute interval) and scan sampling, with individual identification. Wilcoxon test was used to compare the effect of the device insertion on the time spent by cows in each posture and activity. Data are presented as means ± SD. There were statistical differences (P<0.01) in all variables, with reduction of ST (29.28±1.69 min to 26.34±3.91 min, Z=-3.85) and ID (26.48±4.48 min to 23.90±3.80 min, Z=-3.28) and increase of RU (3.42±4.41 min to 5.83±3.84 min, Z=-3.04) and LY (0.66±1.67 min to 3.57±3.69 min, Z=-3.90) when comparing the situation before and after the device insertion. These results suggested that the intravaginal device insertion did not affect cows' behavior in a negative way; rather after the device insertion, the animals seemed to be more relaxed, probably due to an extra habituation to the handling procedures for data collection.

# Authors index

Printed in the United States
by Baker & Taylor Publisher Services